D1179675

083223
Aberdeen College Library

Cultural Industries and the Production of Culture

Over the last few decades there has been a dramatic growth in the importance of non-manufacturing-based forms of production in many Western economies. In addition to the well-documented shift to a service-based economy, many countries have turned increasingly to the production and export of goods and services imbued with high levels of aesthetic and semiotic content (i.e. cultural products). The most obvious examples of this phenomenon are the music, electronic games, television, fashion and media industries, and their emergence as foci of employment growth and economic development, particularly in large global cities. The same trend is associated with radical shifts in global patterns of cultural consumption.

Cultural Industries and the Production of Culture brings together a leading international, multi-disciplinary team of researchers, and presents in an accessible fashion, cutting-edge perspectives on how these industries function, their place in the new economy and how they can be harnessed for urban and regional economic and social development.

Dominic Power is Associate Professor in the Department of Social and Economic Geography, Uppsala University, Sweden, and **Allen J. Scott** is Professor of Policy Studies and Geography at the University of California, Los Angeles.

Routledge studies in international business and the world economy

1. **States and Firms**
   Multinational enterprises in institutional competition
   Razeen Sally

2. **Multinational Restructuring, Internationalization and Small Economies**
   The Swedish case
   Thomas Andersson, Torbjörn Frediksson and Roger Svensson

3. **Foreign Direct Investment and Governments**
   Catalysts for economic restructuring
   Edited by John H. Dunning and Rajneesh Narula

4. **Multinational Investment and Economic Structure**
   Globalization and competitiveness
   Rajneesh Narula

5. **Entrepreneurship in a Global Context**
   Edited by Sue Birley and Ian Macmillan

6. **The Global Structure of Financial Markets**
   An overview
   Edited by Dilip K. Ghosh and Edgar Ortiz

7. **Alliance Capitalism and Global Business**
   John H. Dunning

8. **Multinational Enterprises from the Netherlands**
   Edited by Roger van Hoesel and Rajneesh Narula

9. **Competition, Growth Strategies and the Globalization of Services**
   Real estate advisory services in Japan, Europe and the United States
   Terrence LaPier

10. **European Integration and Foreign Direct Investment in the EU**
    The case of the Korean consumer electronics industry
    Sang Hyup Shin

11 **New Multinational Enterprises from Korea and Taiwan**
Beyond export-led growth
Roger van Hoesel

12 **Competitive Industrial Development in the Age of Information**
The role of co-operation in the technology sector
Edited by Richard J. Braudo and Jeffrey G. MacIntosh

13 **The Global Restructuring of the Steel Industry**
Innovations, institutions and industrial change
Anthony P. D'Costa

14 **Privatization and Liberalization in European Telecommunications**
Comparing Britain, the Netherlands and France
Willem Hulsink

15 **Multinational Corporations**
Emergence and evolution
Paz Estrella Tolentino

16 **Foreign Direct Investment in Emerging Economies**
Corporate strategy and investment behaviour in the Caribbean
Lou Anne A. Barclay

17 **European Integration and Global Corporate Strategies**
Edited by François Chesnais, Grazia Ietto-Gillies and Roberto Simonetti

18 **The Globalization of Corporate R&D**
Implications for innovation systems in host countries
Prasada Reddy

19 **Globalization of Services**
Some implications for theory and practice
Edited by Yair Aharoni and Lilach Nachum

20 **A Century of Foreign Investment in the Third World**
Michael J. Twomey

21 **Global Capitalism at Bay**
John H. Dunning

22 **Foreign Direct Investment**
Research issues
Edited by Bijit Bora

23 **Ford and the Global Strategies of Multinationals**
The North American auto industry
Isabel Studer Noguez

24 **The World Trade Organization Millennium Round**
Freer trade in the next century
Klaus Deutsch and Bernhard Speyer

25 **Consultancy and Innovation**
The business service revolution in Europe
Edited by Peter Wood

26 **Knowledge Economies**
Clusters, learning and co-operative advantage
Philip Cooke

27 **The Role of Resources in Global Competition**
John Fahy

28 **Globalization, Employment and the Workplace**
Diverse impacts
Edited by Yaw A. Debrah and Ian G. Smith

29 **Transnational Corporations**
Fragmentation amidst integration
Grazia Ietto-Gillies

30 **Growth Theory and Growth Policy**
Edited by Harald Hagemann and Stephan Seiter

31 **International Business and the Eclectic Paradigm**
Developing the OLI framework
Edited by John Cantwell and Rajneesh Narula

32 **Regulating Global Trade and the Environment**
Paul Street

33 **Cultural Industries and the Production of Culture**
Edited by Dominic Power and Allen J. Scott

Cultural Industries and the Production of Culture

Edited by Dominic Power and Allen J. Scott

LONDON AND NEW YORK

First published 2004
by Routledge
2 Park Square, Milton Park, Abingdon, Oxon, OX14 4RN

Simultaneously published in the USA and Canada
by Routledge
270 Madison Avenue, New York, NY 10016

Routledge is an imprint of the Taylor & Francis Group

Transferred to Digital Printing 2005

Typeset in Galliard by Wearset Ltd, Boldon, Tyne and Wear

British Library Cataloguing in Publication Data
A catalogue record for this book is available from the British Library

Library of Congress Cataloging in Publication Data
Cultural industries and the production of culture / edited by Dominic Power & Allen J. Scott. — 1st ed.
p. cm. — (Routledge studies in international business and the world economy ; 33)
Includes bibliographical references and index.
1. Arts—Economic aspects. 2. Cultural industries. I. Power, Dominic, 1973– II. Scott, Allen John. III. Series.
NX634.C853 2004
384'.09'045—dc22

2004001844

ISBN 0–415–33101–3

Printed and bound by Antony Rowe Ltd, Eastbourne

Contents

Contributors

Yuko Aoyama (Ph.D. University of California at Berkeley) is an Assistant Professor and Henry J. Leir Faculty Fellow of Geography at Clark University, USA. She specializes in Industrial/Economic Geography with an emphasis on globalization, technological innovation and information society.

Harald Bathelt is Professor of Economic Geography at the Faculty of Geography, Philipps-University of Marburg. His work focuses on innovation, industrial clustering and the socio-economic impacts of regional and industrial change. In his writings, he has also developed a conceptual basis for a relational economic geography.

Nicolas Bautès is completing a Ph.D. in Geography at the University Denis Diderot, Paris 7. He has done extensive research in Rajasthan and Gujarat and works on tourism dynamics and economic development in India and developing countries.

Neil Coe is a Lecturer in the School of Geography at the University of Manchester. His research interests are in the globalization of service activities and, in particular, the film, software and retailing sectors.

John Connell is Professor of Geography, University of Sydney. He is author of *Sydney: Emergence of a World City* (Oxford University Press, 2000). Together with Chris Gibson he has written *Sound Tracks: Popular Music, Identity and Place* (Routledge, 2004).

Louise Crewe is Professor of Human Geography at the University of Nottingham. Her interests lie in the areas of fashion, retailing, shopping and consumption, with particular focus on commodity biographies, value and second-hand exchange. She is the co-author of *Second Hand Cultures* (with Nicky Gregson, Berg, 2003).

Shaun French is currently Lecturer in Economic Geography at the University of Nottingham, having previously studied at the University of Bristol. He has published work on the geographies of business knowledge and praxis.

Chris Gibson is Senior Lecturer in Geography, University of New South Wales. Together with John Connell he has written *Music and Tourism* (Channel View, 2004) and *Deadly Sounds, Deadly Places: Contemporary Aboriginal Music in Australia* (with Peter Dunbar-Hall, UNSW Press, 2004).

Daniel Hallencreutz is Managing Director of a research consultancy company called Intersecta and affiliated to the Centre for Research on Innovation and Industrial Dynamics at Uppsala University. He has a doctorate in Economic Geography from the University of Uppsala and has research interests in the music industry as well as in regional economic development.

Hiro Izushi (Ph.D. University of California at Berkeley) is a Senior Lecturer at the Centre for Local Economic Development, Coventry Business School, U.K. He specializes in technological innovation with an emphasis on knowledge management, technology transfer, and technical change and economic growth.

Jennifer Johns is currently completing a Ph.D. at the School of Geography, University of Manchester, entitled "Tracing the Connections: Manchester's Film and Television Production Networks."

Andrew Leyshon is Professor of Economic Geography at the University of Nottingham. In addition to investigating the origins and impacts of e-commerce, he is also currently researching the formation of ecologies of retail financial services. He is the author of *Money/Space: Geographies of Monetary Transformation* (with Nigel Thrift, Routledge, 1997), and co-editor of *The Place of Music* (with David Matless and George Revill, Guilford, 1998) and *Alternative Economic Spaces* (with Roger Lee and Colin Williams, Sage, 2003).

Angela McRobbie is Professor of Communications at Goldsmiths College London. She is the author of several books on gender, popular culture, and "subcultural entrepreneurs." These include *Feminism and Youth Culture* (Palgrave, 2nd edn 2000), *British Fashion Design* (Routledge, 1998) and *In the Culture Society* (Routledge, 1999). Her recent research is on the experiential aspects of working in the culture industries in London and Berlin.

Justin O'Connor has been Director of the Manchester Institute of Popular Culture MIPC since 1995 and is also Reader in Sociology at the Manchester Metropolitan University. He has research interests and has published extensively in the areas of urban cultures and lifestyles, urban regeneration, cultural policy and city imaging, cultural industries and popular culture. His book, *Cultural Industries and the City*, will appear in 2005.

Jane Pollard is a Senior Lecturer in the Centre for Urban and Regional Development Studies at the University of Newcastle upon Tyne. Her research interests include geographies of money and finance, and the role of different financial intermediaries in regional economic development, post colonial economic geographies, and the changing nature and practices of economic geography.

Dominic Power is an Associate Professor in Economic Geography at Uppsala University. His research is concerned with regional and industrial competitiveness, and the workings of the cultural industries, in particular the music and design industries.

Andy Pratt is a Senior Lecturer in Human Geography at the London School of Economics and Politics Science. His particular concerns are with macro-level dynamics in the cultural industries and policy, with the constituent industries, and with the process of organization, especially those of localization.

Norma Rantisi is an Assistant Professor in the Department of Geography, Planning and Environment at Concordia University, Montréal, Québec, Canada. Her research interests focus on the restructuring of mature manufacturing industries and the cultural economy of cities. She has recently published articles on the evolution and design innovation process of the fashion industry in New York City.

Walter Santagata is Professor of Economics at Turin University. He is head of the Department of Economics and director of the course "Cultural Projects for Development," organized by I.T.C.-I.L.O. He has published extensively on cultural economics.

Allen J. Scott is Professor with joint appointments in the Department of Geography and the Department of Policy Studies at UCLA. He was awarded the Vautrin Lud Prize in 2003. His most recently published books are *The Cultural Economy of Cities* (Sage, 2000) and *On Hollywood* (Princeton University Press, 2004).

Nigel Thrift is Head of the Division of Life and Environmental Sciences at the University of Oxford. His main research interests are in the history of time consciousness, geographies of capitalism, cultural economy, spaces of cognitive assistance and nonrepresentational theory. Recent books have included *Cities* (with Ash Amin, Polity, 2002), *The Blackwell Cultural Economy Reader* (co-edited with Ash Amin, Blackwell, 2004), and *Patterned Ground* (co-edited with Stephan Harrison and Steve Pile, Reaktion, 2004).

Elodie Valette has a Ph.D. in Geography. She works on social and economic innovation processes in rural and peri-urban territories in France and other Western countries.

Peter Webb studied Politics and Sociology at the Universities of the West of England and Bristol. He is currently a Lecturer in Sociology at the University of Birmingham. Prior to this he was Research Fellow at the University of Bristol. He has published work on popular music and social theory.

Part I

Introduction

1 A prelude to cultural industries and the production of culture

Dominic Power and Allen J. Scott

The rise of the cultural economy

Since the early 1980s, a so-called new economy has steadily risen to prominence as a focus of employment and output growth in virtually all the major capitalist societies. This new economy is represented primarily by sectors such as high-technology manufacturing, neo-artisanal consumer products and diverse services, all of which have a propensity to take organizational shape as complex value-added networks. The operating features of these networks rest on their doubly-faceted character as congeries of many small firms together with more restricted cohorts of large establishments, the latter, more often than not, forming units within even larger corporate conglomerates. At the same time, these networks are much given to high levels of organizational and technological flexibility, transactions-intensive inter-firm relations and the production of design-intensive outputs.

One of the most important segments of this new economy comprises a group of industries that can be loosely identified as suppliers of cultural products (Scott, 2000). The rapid growth and spread of these industries in recent decades is a reflection of the increasing convergence that is occurring in modern society between the economic order on the one hand and systems of cultural expression on the other hand (Lash and Urry, 1994). These industries produce an enormous and ever-increasing range of outputs. Examples of these, all of which figure prominently in the present book, are jewelry (Pollard), music (French *et al.*, Gibson and Connell, Power and Hallencreutz), video games (Aoyama and Izushi), film and television (Coe and Johns), new media (Bathelt), fashion design (French *et al.*, Santagata, Rantisi) and the visual arts (McRobbie, Valette and Bautès).

The industries that make up the contemporary cultural economy are bound together as an object of study by three important common features. First, they are all concerned in one way or another with the creation of products whose value rests primarily on their symbolic content and the ways in which it stimulates the experiential reactions of consumers (Bourdieu, 1971; Pine and Gilmore, 1999). Second, they are generally subject to the effects of (Ernst) Engels' Law, which suggests that as disposable income expands so consumption of non-essential or luxury products will rise at a disproportionately higher rate. Hence, the richer the country, the higher expenditure on cultural products will be as a fraction of families' budgets. Third, firms in cultural-products industries are subject to competitive and organizational pressures such that they frequently agglomerate together in dense specialized clusters or industrial districts, while their products circulate with increasing ease on global markets.

It must be stressed at once that there can be no hard and fast line separating industries that specialize in purely cultural products from those whose outputs are purely utilitarian. On the contrary, there is a more or less unbroken continuum of sectors ranging from, say, motion pictures or recorded music at the one extreme, through an intermediate series of sectors whose outputs are varying composites of the cultural and the utilitarian (such as shoes, eye-glasses or cars), to, say, cement or petroleum products at the other extreme. At the same time, one of the peculiarities of modern capitalism is that the cultural economy continues to expand at a rapid pace not only as a function of the growth of discretionary income, but also as an expression of the incursion of sign-value into ever-widening spheres of productive activity at large as firms seek to intensify the design content, styling and quality of their outputs in the endless search for competitive advantage.

Technology, organization and work in the cultural economy

Over much of the twentieth century, the leading edges of economic development and growth were largely identifiable with sectors characterized by varying degrees of mass production, as expressed in large-scale machine systems and a persistent drive to product standardization and cost cutting. Throughout the mass-production era, the dominant sectors evolved through a succession of technological and organizational changes focused above all on process routinization and the search for internal economies of scale. These features are not especially conducive to the injection of high levels of aesthetic and semiotic content into final products. Indeed, in the 1930s and 1940s many commentators – with adherents of the Frankfurt School (e.g. Adorno, 1991; Horkheimer, 1947) being among the most vocal – expressed grave misgivings about the steady incursion of industrial methods into the sphere of the cultural economy and the concomitant tendency for complex social and emotive content to be evacuated from forms of popular cultural production. These misgivings were by no means out of place in a context where much of commercial culture was focused on an extremely narrow approach to entertainment and distraction, and in which the powerful forces of the nation-state and nationalism were bent in significant ways on creating mass proletarian societies. The specific problems raised by the Frankfurt School in regard to popular commercial culture have in certain respects lost some of their urgency as the economic and political bases of mass production have given way before the changes ushered in over the late 1970s and early 1980s, when the new economy started its ascent. This is not to say that the contemporary cultural economy is not associated with a number of serious social and political predicaments. But it is also the case that as commercial cultural production and consumption have evolved in the major capitalist societies over the last few decades, so our aesthetic and ideological judgments about their underlying meanings have tended to shift. The rise of post-modern social and cultural theory is one important expression of this development (cf. Jameson, 1992)

In contrast to mass machinofacture, new economy sectors tend to be composed of relatively disarticulatable production processes, as represented, for example, by various kinds of computerized and digitized technologies. Equally, and nowhere more than in segments of the new cultural economy, production is often quite labor-intensive, for despite the widespread use of electronic technologies, it also tends to make heavy demands on both the brain-power and handiwork of the labor force. Like most other sectors that make up the new economy, cultural-products industries are typically composed of swarms of small producers (with low entry and exit costs), complemented by many fewer numbers of large establishments. Small producers in the cultural economy are frequently marked by neo-artisanal forms of production, or, in a more or less equivalent phrase, to flexible specialization, meaning that they concentrate on making particular categories of products (clothing, films, games, etc.) but where the design specifications of each batch of products change repeatedly. Large firms in the cultural economy occasionally tend toward mass production (which would generally signify a diminution of symbolic function in final outputs), but are increasingly prone to organization along the lines of "systems houses" (Scott, 2002). The latter term is used in the world of high-technology industry to signify an establishment whose products are relatively small in number over some fairly extended period of time, and where each unit of output represents huge inputs of capital and/or labor. The major Hollywood movie studios are classic cases of systems houses. Other examples of the same phenomenon – or close relatives – are large magazine publishers (but not printers), electronic games producers, television network operators and, to a lesser degree, leading fashion houses. These large producers are of particular importance in the cultural economy because they so frequently act as the hubs of wider production networks incorporating many smaller firms. Equally, and above all in the entertainment industry, they play a critical part in the financing and distribution of much independent production. In addition, large producers right across the cultural economy are increasingly subject to incorporation into the organizational structures and spheres of influence of giant multinational conglomerates through which they tap into huge financial resources and marketing capacities. While these giant firms are absolutely central to the cultural economy, it is also important to note that some aspects of their power and reach may currently be under threat. There is already evidence of this occurring in the music and film industries where many new possibilities for disintermediation and distribution have been brought about by recent technological advances (see the chapter by French *et al.*). Indeed, one can imagine that at least some segments of the cultural economy – resting as it does on fluid and unpredictable trends and hard-to-protect intellectual property – may be entering into a new phase of development marked by yet more intense competition and reduced levels of oligopolistic power.

The actual work of production in the cultural economy is typically carried out within shifting networks of specialized but complementary firms. Such networks assume different forms, ranging from heterarchic webs of small establishments to more hierarchical structures in which the activities of groups of establishments are coordinated by a dominating central unit, with every possible variation between these two extreme cases. As analysts like Caves (2000) have observed, much of the cultural economy can be described as conforming to a contractual and transactional model of production. The same model extends to the employment relation, with part-time, temporary and freelance work being particularly prevalent. The instabilities associated with this state of affairs often lead to intensive social networking activities among skilled creative workers as a means of keeping abreast of current labor-market trends and opportunities and of finding collaborators, customers and employers (Scott, 1998). Within the firm, these same workers are often incorporated into project-oriented teams, a form of work organization that is rapidly becoming the preferred means of managing internal divisions of labor in the more innovative segments of the modern cultural economy (Grabher, 2002). By contrast, in sectors such as clothing or furniture, where low-wage manual operators usually account for a high proportion of total employment, piece-work and sweatshop conditions are more apt to be the prevailing modes of incorporating workers into the production process, though these sectors also have high-wage, high-skill segments.

These features of the new economy in general (and the modern cultural economy in particular) differentiate it quite markedly from the older model of mass production. In contrast to what was often seen as the dispiriting and endless uniformity of the outputs that flowed from the mass-production system, the new economy is marked by extremely high levels of product variety. As a corollary, the new economy is associated with a major transformation of market structures, with monopolistic competition *à la* Chamberlin (1933) becoming increasingly the norm. Chamberlinian competition, which resembles in some respects imperfect competition as formulated by Robinson (1933), is based on the notion that distinctive market distortions appear when producers have strongly-developed firm-specific characteristics. Under a regime of monopolistic competition there may be many individual firms all making a particular class of products, but each firm's output also has unique attributes (design, place-specific associations, brand, etc.) that can only be imitated by other firms in the form of inferior reproductions, or can at best be copied only after a significant time lag. The increasing importance of cultural and symbolic content in contemporary patterns of consumption means that monopolistic competition has become an ever more feasible option for firms throughout the new economy. The constant rebranding and repackaging characteristic of product markets today is helping to usher in an economic system where even small firms can sometimes vie with goliaths in the creation of virtual product monopolies.

The chapters by Rantisi and Santagata highlight some of the critical issues here, and, in particular, the important interconnections that run between innovative performance, fashion and styling, and market dynamics in today's cultural economy.

Place and production in the cultural economy

Cultural-products industries with attributes like these almost always operate most effectively when the individual establishments that make up them exhibit at least some degree of locational agglomeration. In fact, the most common leitmotif that connects the following chapters together is their emphasis in one way or another upon the persistent tendency of producers in the cultural economy to cluster together in geographic space.

This tendency follows at once from the economic efficiencies that can be obtained when many different interrelated firms and workers lie in close proximity to one another so that their complex interactions are tightly circumscribed in space and time. But agglomeration also occurs for reasons other than economic efficiency in the narrow sense. It is also partly a result of the learning processes and innovative energies that are unleashed from time to time in industrial clusters as information, opinions, cultural sensibilities, and so on, are transmitted through them, and these processes are usually especially strong in cases where transactional intensity is high. Moreover, outputs that are rich in information, sign value and social meaning are particularly sensitive to the influence of geographic context and creative milieu. This point is strongly echoed in the chapters by Rantisi, Bautès and Valette, McRobbie, Bathelt, Power and Hallencreutz, and Pollard. In the same way, Molotch (1996) has argued that agglomerations of design-intensive industries acquire place-specific competitive advantages by reason of local cultural symbologies that become congealed in their products, and that imbue them with authentic character. This intensifies the play of Chamberlinian competition in the cultural economy because monopolistic assets now not only emerge from the productive strategies of individual firms, but also from their wider geographic milieu.

The association between place and product in the cultural industries is often so strong that it constitutes a significant element of firms' successes on wider markets. Place-related markers, indeed, may become brands in themselves that firms can exploit to increase their competitive positions, as exemplified by the cases of Parisian fashions, Jamaican reggae, Danish furniture or Italian shoes. Successful cultural-products agglomerations, as well, are irresistible to talented individuals who flock in from every distant corner in pursuit of professional fulfillment, in a process that Menger (1993) has referred to as "artistic gravitation." The gravitational forces exerted by agglomerations of creative industries and their associated cohorts of workers are often of considerable power. In the present book, McRobbie writes of artists gathering together in creative hubs in London, and Rantisi alludes to the pull of New York's fashion industry on clothing designers. Gravitational forces such as these mean that the labor pools of dynamic agglomerations are constantly being replenished by selective in-migration of workers who are already predisposed to high levels of job performance in the local area. Local supplies of relevant skills and worker sensibilities are further augmented by the specialized educational and training institutions that typically spring into being in productive agglomerations.

These remarks indicate that a tight interweaving of place and production system is one of the essential features of the new cultural economy of capitalism. In cultural-products industries, as never before, the wider urban and social environment and the apparatus of production merge together in potent synergistic combinations. Some of the most advanced expressions of this propensity can be observed in great world cities like New York, Los Angeles, Paris, London or Tokyo. Certain districts in these cities are typified by a more or less organic continuity between their place-specific settings (as expressed in streetscapes, shopping and entertainment facilities, and architectural background), their social and cultural infrastructures (museums, art galleries, theaters, and so on), and their industrial vocations (advertising, graphic design, audiovisual services, publishing or fashion clothing, to mention only a few). Such cities often seek to promote this continuity by consciously re-organizing critical sections of their internal spaces like theme parks and movie sets, as exemplified by Times Square in New York, The Grove in Los Angeles or Potsdamer Platz in Berlin (Zukin, 1995). In a city like Las Vegas, the urban environment, the production system and the world of the consumer are all so tightly interwoven as to form a virtually indivisible unity. The city of work and the city of leisure increasingly interpenetrate one another in today's world.

Cultural industries and local economic development policy

Cultural-products industries are growing rapidly; they tend (though not always) to be environmentally-friendly; and they frequently (though again not always) employ high-skill, high-wage, creative workers. Cultural-products industries also generate positive externalities in so far as they contribute to the quality of life in the places where they congregate and enhance the image and prestige of the local area. Moreover, as noted above, they tend to be highly localized and often place-bound industries. This fact has made them increasingly attractive to policy-makers intent on finding new solutions to problems of urban redevelopment and local economic performance.

A sort of first-generation approach to the systematic deployment of cultural assets in the quest for local economic growth can be found in the aggressive place-marketing pursued by many municipal authorities since the early 1980s (Philo and Kearns, 1993). This activity is often based on a local heritage of historical or artistic resources, but it also assumes the guise of energetic redevelopment programs. One of the most remarkable instances of the remaking and marketing of place in recent years is furnished by the Guggenheim Museum in Bilbao, an initiative that has turned an old and stagnant industrial area into a world-renowned tourist center and a new focus of inward investment. Other examples of similar phenomena can be found in Seville's Expo or in the promotion of tourism in a number of old British industrial cities (notably Liverpool and Manchester) on the basis of their roles as popular music centers. A striking example can also be found in the case of the Ruhr region where the repackaging of an old industrial landscape has helped to spark off major new cultural developments (Gnad 2000). O'Connor's chapter on the cultural economy of St. Petersburg provides further insights into the role of place-marketing in local economic development. One of the dilemmas of this approach, as highlighted by O'Connor, is the mismatch that sometimes occurs between the exigencies of the tourist trade and local cultural aspirations and attitudes.

Of late years, an alternative (or, rather, a complementary) second generation of policy approaches has come increasingly under the scrutiny of local authorities. In this instance, the objective is less the construction or redevelopment of facilities that will entice visitors to flock into a given center, as it is to stimulate the formation of localized complexes of cultural industries that will then export their outputs far and wide. The evolving Leipzig media cluster, described in Bathelt's chapter, is one example. In Leipzig, policy-makers have sought to (re)build a network of media firms that are strongly anchored to the local area through tightly-knit firm-to-firm links and institutional infrastructures. Equally, Pollard shows how the jewelry industry of Birmingham is being revitalized – though not always with unqualified success – by concerted attempts to upgrade skills and product quality within an intricate social division of labor. Even more than in the case of place-marketing approaches, this alternative line of attack is critically dependent on a clear understanding of the logic and dynamics of the agglomeration processes that shape much of the geography of the modern cultural economy.

For any given agglomeration, the essential first task that policy-makers must face is to map out the collective order of the local economy along with the multiple sources of the increasing-returns effects that invariably emanate from its inner workings. This in itself is a difficult task due both to the problems of defining just where the cultural economy begins and ends, and to the intangible nature of many of the phenomena that lie at the core of localized competitive advantages (see Pratt's chapter). That said, it is this collective order more than anything else that presents possibilities for meaningful and effective policy intervention in any given agglomeration. Blunt top-down approaches focused on directive planning are unlikely in and of themselves to accomplish much at the local scale, except in special circumstances. In terms of costs and benefits and general workability, the most successful types of policies will as a general rule be those that concentrate on the character of localized external economies of scale and scope as public or quasi-public goods. The point here is both to stimulate the formation of useful agglomeration effects that would otherwise be undersupplied or dissipated in the local economy, and to ensure that existing externalities are not subject to severe misallocation as a result of market failure. Finely tuned bottom-up measures are essential in situations like this.

Policy-makers thus need to pay attention to three main ways of promoting collective competitive advantage, which, on the basis of the modern theory of industrial districts can be identified as (a) the building of collaborative inter-firm relations in order to mobilize latent synergies, (b) the organization of efficient, high-skill local labor markets and (c) the potentiation of local industrial creativity and innovation (cf. Scott, 2000; Storper, 1997). The specific means by which these broad objectives can be pursued are many and various depending on empirical circumstances, but basic institution-building in order to internalize latent and actual externalities within competent agencies and to coordinate disparate groups of actors is likely to be of major importance. Complementary lines of attack involve approaches such as the initiation of labor-training programs, creating centers for the encouragement of technological upgrading or design excellence, organizing exhibitions and export drives, and so on, as well as socio-juridical interventions like dealing with threats to the reputation of local product quality due to free rider problems (especially in tourist resorts), or helping to protect communal intellectual property. In addition, appropriately structured private–public partnerships could conceivably function as a vehicle for generating early warning signals as and when the local economy appears to be in danger of locking into a low-level equilibrium due to adverse path-dependent selection dynamics. The latter problem is especially apt to make its appearance in localized production systems because the complex, structured interdependencies within them often give rise to long-run developmental rigidities.

While most development based on cultural products industries will in all likelihood continue to occur in the world's richest countries, a number of low- and middle-income countries are finding that they too are able to participate in various ways in the new cultural economy, sometimes on the basis of traditional industries and cultures (such as the miniature paintings of Rajasthan described by Bautès and Valette, or the music of Jamaica that Power and Hallencreutz deal with in their chapter). Even old and economically depressed industrial areas, as we have seen, can occasionally turn their fortunes around by means of well-planned cultural initiatives.

To be sure, the notion of the cultural economy as a source of economic development is still something of a novelty, and much further reflection is required if we are to understand and exploit its full potential while simultaneously maintaining a clear grasp of its practical limitations. In any case, an accelerating convergence between the economic and the cultural is currently occurring in modern life, and is bringing in its train new kinds of urban and regional outcomes and opening up new opportunities for policy-makers to raise local levels of income, employment and social well-being.

The global connection

In spite of the predisposition of firms in particular cultural-products industries to locate in close mutual proximity, their outputs flow with relative ease across national borders and are a steadily rising component of international trade. The international flow of cultural goods and services is reinforced by the operations of transnational media conglomerates whose main competitive strategy appears increasingly to be focused on the creation of worldwide blockbuster products, as exemplified dramatically by the market offerings of major firms in the Hollywood film industry. At the same time, with ever greater global interconnectedness many different cultural styles and genres become accessible to far-flung consumers so that highly specialized niche markets are also proliferating alongside the blockbuster markets in which major corporations largely participate. Consumers in particular niches (such as computer games) often assert their collective identity by means of web-based forms of intercommunication, thereby increasing their visibility as distinctive market fractions. With the further development of new electronic distribution technologies (such as the Internet and mobile telephones) for cultural products, the process of globalization will assuredly accelerate, at least for cases where digitization of final outputs is feasible.

Observe that globalization in the sense indicated does not necessarily lead to the locational dispersal of production itself. On the contrary, globalization *qua* spatial fluidity of end products helps to accentuate agglomeration because it leads to rising exports combined with expansion of localized production activities. Concomitant widening and deepening of the social division of labor at the point of production then helps to intensify clustering because it generates increased positive externalities. Locational agglomeration and globalization, in short, are complementary processes under specifiable social and economic circumstances. That said, the falling external transactions costs associated with globalization will sometimes undermine agglomeration from the other end, as it were, by making it feasible for some kinds of production to move to alternative locations. It is now increasingly possible for given activities that could not previously escape the centripetal forces of agglomeration to decentralize to cheap labor sites. This may result in a wide dispersal of certain types of production units, such as plants processing CD-ROMs for the recording industry, or teams of motion-picture workers engaged in location shooting. In other instances, it is expressed in the formation of alternative clusters or satellite production locations, as illustrated by the sound stages and associated facilities that have come into existence in Toronto and Vancouver (Canada) and Sydney (Australia) in order to serve film-production companies in Hollywood. This is a topic that Coe and Johns take up in their chapter.

The overall outcome of these competing spatial tensions in the modern cultural economy is a widening global constellation of production centers. The logic of agglomeration and increasing-returns effects suggests that one premier global center will occasionally emerge in any given sector, but even in the case of the international motion-picture industry, which is overwhelmingly dominated by Hollywood, it can be plausibly argued (above all in a world of monopolistic competition) that multiple production centers will continue to exist if not flourish. At any rate, the scenario of thriving multiple production centers is defendable provided that policy-makers and others pay due attention to fostering their associated distribution and marketing systems. Two of the following chapters – Gibson and Connell's chapter on the Australian indigenous popular music industry and Power and Hallencreutz's comparison of the Swedish and Jamaican music industries – show that successful development of localized cultural-products industries depends intimately on being integrated, in the right ways and at firm level, into global circuits of production and consumption. Large multinational corporations play a decisive role across this entire functional and spatial field of economic activity, both in coordinating local production networks and in ensuring that their products are projected onto wider markets. This remark, by the way, should not induce us to neglect the fact that small independent firms continue to occupy an important place in almost all cultural-products agglomerations. In the past, multinationals based in the United States have led the race to command global markets for nearly all types of cultural products, but producers from other countries are now entering the fray in ever greater numbers, even in the media sectors that have hitherto been considered as the privileged preserve of North American firms. In the same way, different cultural-products industrial agglomerations around the world are increasingly caught up with one another in global webs of co-productions and creative partnerships. Bathelt *et al.* (2004) point to the circumstance that no localized group of firms can nowadays be completely self-sufficient in terms of state-of-the-art knowledge-creation, and that world-wide inter-agglomeration networks are an increasingly vital element of any individual agglomeration's performance. Concomitantly, global productive alliances and joint ventures are surging to the fore in the modern cultural economy, drawing on the specific competitive advantages of diverse clusters, but without necessarily compromising the underlying force of agglomeration itself. Furthermore, the profitability of cultural products is more and more dependent upon the establishment of an effective international regime of intellectual property rights ensuring the protection of particular artists, firms and localities from product piracy and imitation.

Once all of this has been said, the advent of a new cultural economy and the flow of its outputs through circuits of international commerce have not always been attended by benign results. This situation has in fact led to numerous political collisions over issues of trade and culture. One of the more outstanding instances of this propensity is the clash that occurred between the United States and Europe over trade in audiovisual products at the time of the GATT (General Agreement on Trade and Tariffs)[1] negotiations in 1993. Notwithstanding such notes of dissonance, we seem to be moving steadily into a world that is becoming more and more cosmopolitan and eclectic in its modes of cultural consumption. Certainly for consumers in the more economically-advanced parts of the world, the standard American staples (from fast food to films) are now but one element of an ever widening palette of cultural offerings comprising Latin American telenovelas, Japanese comic books, Hong Kong kung fu movies, indigenous Australian music, London fashions, African literature, Balinese tourist resorts, Chilean wines, Mexican cuisine and untold other exotic fare. This trend is in significant degree both an outcome of and a contributing factor to the recent, if still incipient, advent of a multifaceted and extensive global system of cultural-products agglomerations. In view of these comments, and despite the corrosive effects of commercial culture on many more traditional forms of cultural production, globalization does not appear to be leading to cultural uniformity so much as it is to increases in the variety of options open to individual consumers.

A concluding comment

The cultural economy now accounts for substantial shares of income and employment in a wide range of countries (García *et al.*, 2003; Power, 2002, 2003; Pratt, 1997; Scott, 1996). By the same token, it offers important opportunities to policy-makers in regard to local economic development.

In this discussion, we have concentrated almost entirely on the economic side of this equation, but the observations offered here raise an equally important set of issues concerning cultural politics, not only in regard to trade, but also, and more significantly, in regard to matters of human diversity, growth and development generally. As cryptic as this remark may be, it opens up a vast terrain of debate about the qualitative meaning of the overarching system of cultural consumption that is being ushered into existence by the trends and processes discussed in this book. The goods and services that sustain this system are to ever-increasing degrees fabricated within production networks organized according to the logic of capitalist enterprise and concentrated within far-flung industrial clusters. One important effect of this condition is the increasing diversity of the cultural products that now flow through global markets. Another is the ephemerality and attenuated symbolic intensity that characterize significant segments of commercialized cultural production today. Yet another is the enormous disparities between different groups of individuals in different societies in regard to their command of cultural resources and forms of self-expression. A progressive (but still largely prospective) cultural politics attuned to such issues will no doubt take as one of its main objectives the further encouragement of product diversity in the global cultural economy while seeking to enhance the critical capacities of consumers. Indeed an underlying premise throughout this book is the positive possibilities – economic and social – that rise into view as cultural-products industries come to account for a greater and greater share of the modern economy. Harnessed in the right way these industries can work not only to promote economic growth and prosperity but also to enrich people's everyday lives and experiences. There is always, however, a tense interplay between the commercial imperatives of the cultural economy and the actual meanings embedded in its products. The great danger, in the absence of some effective counter-weight to the supply side of this equation is that many of the more progressive potentialities of the modern cultural economy may remain severely underdeveloped.

The chapters that follow pursue the themes raised in this introductory essay in much more detail. Taken as a whole, these chapters greatly advance the current state of knowledge about the new cultural economy of capitalism, both as a socio-economic phenomenon and as a focus of multiple normative concerns. It is our hope that the book will provide a platform for further intensive research into the important analytical and policy issues raised by the recent and world-wide efflorescence of cultural industries and products.

Note

1 Now the WTO (World Trade Organization).

References

Adorno, T. (1991) *The Culture Industry: Selected Essays on Mass Culture*, London: Routledge.

Bathelt, H., Malmberg, A. and Maskell P. (2004) "Clusters and knowledge: local buzz, global pipelines and the process of knowledge creation," *Progress in Human Geography*, 28: 54–79.

Bourdieu, P. (1971) "Le marché des biens symboliques," *L'Année Sociologique*, 22: 49–126.

Caves, R. E. (2000) *Creative Industries: Contacts between Art and Commerce*, Cambridge, MA: Harvard University Press.

Chamberlin, E. (1933) *The Theory of Monopolistic Competition*, Cambridge, MA: Harvard University Press.

García, M. I., Fernández, Y. and Zofío, J. L. (2003) "The economic dimension of the culture and leisure industry in Spain: national, sectoral and regional analysis," *Journal of Cultural Economics*, 27: 9–30.

Gnad, F. (2000) "Regional promotion strategies for the culture industries in the Ruhr area," in F. Gnad and J. Siegmann (eds) *Culture Industries in Europe: Regional Development Concepts for Private-Sector Cultural Production and Services*, Düsseldorf, Ministry for Economics and Business, Technology and Transport of the State of North Rhine-Westphalia, and the Ministry for Employment, Social Affairs and Urban Development, Culture and Sports of the State of North Rhine-Westphalia.

Grabher, G. (2002) "Cool projects, boring institutions: temporary collaboration in social context," *Regional Studies*, 36: 205–14.

Horkheimer, M. (1947) *The Eclipse of Reason*, New York: Oxford University Press.

Jameson, F. (1992) *Postmodernism, or, the Cultural Logic of Late Capitalism*, Durham, NC: Duke University Press.

Lash, S. and J. Urry (1994) *Economies of Signs and Space*, London/Thousand Oaks, CA: Sage.

Menger, P. M. (1993) "L'hégémonie parisienne: économie et politique de la gravitation artistique," *Annales: Economies, Sociétés, Civilisations*, 6: 1565–600.

Molotch, H. (1996) "LA as design product: how art works in a regional economy," in A. J. Scott and E. W. Soja (eds) *The City: Los Angeles and Urban Theory at the End of the Twentieth Century*, Berkeley and Los Angeles: University of California Press.

Philo, C. and Kearns, G. (1993) "Culture, history, capital: a critical introduction to the selling of places," in G. Kearns and C. Philo (eds) *Selling Places: The City as Cultural Capital, Past and Present*, Oxford: Pergamon Press.

Pine, B. and Gilmore, J. (1999) *The Experience Economy: Work is Theatre and Every Business a Stage*, Boston: Harvard Business School Press.

Power, D. (2002) "Cultural industries in Sweden: an assessment of their place in the Swedish economy," *Economic Geography*, 78: 103–27.

—— (2003) "The Nordic cultural industries: a cross-national assessment of the place of the cultural industries in Denmark, Finland, Norway and Sweden," *Geografiska Annaler B*, 85: 167–80.

Pratt, A. C. (1997) "The cultural industries production system: a case study of employment change in Britain, 1984–91," *Environment and Planning A*, 29: 1953–74.

Robinson, J. (1933) *The Economics of Imperfect Competition*, London: Macmillan.

Scott, A. J. (1996) "The craft, fashion, and cultural products industries of Los Angeles: competitive dynamics and policy dilemmas in a multi-sectoral image-producing complex," *Annals of the Association of American Geographers*, 86: 306–23

—— (1998) "Multimedia and digital visual effects: an emerging local labor market," *Monthly Labor Review*, 121: 30–8.

—— (2000) *The Cultural Economy of Cities: Essays on the Geography of Image-Producing Industries*, London: Sage.

—— (2002) "A new map of Hollywood: the production and distribution of American motion pictures," *Regional Studies*, 36: 957–75.

Storper, M. (1997) *The Regional World: Territorial Development in a Global Economy*, New York: Guilford Press.

Zukin, S. (1995) *The Cultures of Cities*, Oxford: Blackwell.

Part II

Trends and opportunities in the cultural economy

2 Mapping the cultural industries

Regionalization; the example of South East England

Andy C. Pratt

Introduction

The aim of this chapter is to follow attempts to both define and develop empirical measures that may act as an evidential base for policy making in the area of the cultural industries. Specifically, the chapter will focus on the shift from a national to a regional basis of cultural industries policy making. The first step in this process in the United Kingdom (U.K.) was the production of the *Creative Industries Mapping Document* (DCMS, 1998). This document sought to use secondary sources to record the contribution of the cultural industries[1] to the whole U.K. economy. Remarkably, few states[2] had thought of carrying out such an exercise before; and certainly none were prepared to publicize the outcomes so widely. The four headline statistics from this report were as follows: that the cultural industries employed close to 1.4 million persons, which represented 5 percent of the total U.K. workforce at the time; revenues from the cultural industries was in the excess of £60bn; they contributed £7.5bn to export earnings (excluding intellectual property); and value added (net of inputs) was £25bn, which significantly was 4 percent of U.K. GDP, and in excess of any (traditional) manufacturing industry.

This reported success of the cultural industries was a surprise to both policy makers and the public for two interrelated reasons. First, data had not been systematically collected previously. Second, the cultural sector as a whole, and particularly its commercial element, is a relatively new one that has grown very quickly. Policy makers had overlooked the cultural industries because traditional business census classifications were insufficiently calibrated to identify them. It is only by significant data manipulation that the contribution of the sector can be separated out (see Pratt, 1997).

The significance of this new analysis cannot be underestimated; for policy makers it is as if suddenly a successful new industry has arrived from nowhere. Although the constituent industries (film, television, advertising, etc.) are widely recognized, previously they have been seen either as part of the state-supported sector, or viewed as somewhat peripheral to the "real" economy (except in the United States, see Siwek, 2002). The new data has shown this not to be the case. The U.K. report was widely circulated and it attracted a considerable amount of attention – many governments went to look at their own cultural sector, or turned their attention to building, or attracting the cultural industries to their economies. Perhaps the most important element of the U.K.'s promotion of its cultural industries was the recognition of the economic dimension of culture, which has led to a re-thinking of cultural policy. This economic image of the cultural sector contrasts sharply with traditional conceptualizations of cultural policy as either heritage management or as a humanist ideal. It is for these reasons that the U.K. has become a model of new cultural policy that picks a path between, for example, the dirigiste model of France, and the *laissez-faire* approach of the United States.

The focus of this chapter is to explore "what happened next" to the cultural industries agenda in the U.K. The simple answer is "regionalization." Despite the inclusion of "mapping" in the title there are no maps in the U.K. report; moreover, there was no attempt to address regional or local variations, or to benchmark findings against those of other nation states. In the period since 2000 the U.K. government has begun a process of devolution and the creation of a form of regional governance. The second *Mapping Document* (DCMS, 2001) acknowledged this concern, albeit with only a cursory review of regional initiatives and, significantly, with no substantive data. This chapter plots the emergence of a regional cultural industries agenda; specifically, the problems of, and need to establish, meaningful regional datasets that might both make visible the regional status of the creative industries, as well as informing prospective policy. The chapter begins by sketching in the "regionalization of culture" that has taken place in the U.K. since 2000. Second, it outlines the problems of, and some solutions to, the legitimation issue: the establishment of an evidential base for the creative industries. Finally, it provides an illustration of regional cultural data analysis in South East England.

The regionalization of the creative industries

A notable theme of the post-1997 Labour administration has been a reform of regional governance in the U.K. On one hand, the devolution of the nations (Scotland, Wales and Northern Ireland) and, on the other, the empowerment of the English regions. In practice, this process has not been a simple or stress-free policy agenda; however, it is not the place to discuss these issues here (see Tomaney, 1999, 2000). The objective here is to locate the cultural industries agenda within this process; specifically, the challenge of how to "re-size" cultural industries policy from a national to a regional scale.

The eight English Regional Development Agencies were born in 1999. Each agency was required to develop a strategy with the overall objective of improving economic performance.[3] All government departments had to consider the regional dimensions of their activities. For some, such as the Department of Culture, Media and Sport (DCMS), this required some structural changes. The problem was that the DCMS has no proper regional structure: it works through the regional Government Offices, which pre-dated Regional Development

Agencies, but it is Whitehall focused. The Arts Council, the traditional home of arts and cultural policy, is what is known in policy debates as an "arms-length" body: it is directly funded by the DCMS but its policy-making function is independent. The Arts Council did have a regional structure, in the shape of the Regional Arts Boards. However, the relationship between the Arts Council and its regions was strained: there was a perception in the regions of central control, and a metropolitan basis for funding allocation; the reorganization of the Arts Council and the creation of the new Regional Arts Councils (2003) was an attempt to address these issues.

The problem for the DCMS was how to step into this complex institutional matrix and to respond to both the regional agenda and the growing tension between the Department's economic and cultural constituencies. This latter point had become salient as the cultural industries, with their commercial and economic profile, were considered by many in the arts establishment to be in opposition, in aim and purpose, to "the arts." In 2000 the DCMS established Regional Cultural Consortia – the role of these new agencies was to work with Regional Arts Boards and the Regional Development Agencies to provide a "joined up" approach to the delivery of cultural services.

The growing power of the Regional Development Agencies, and their economic agenda, quickly drew the Regional Cultural Consortia to focus their attentions on the cultural industries. In no small measure this was because the Regional Development Agencies were required to take into account regional cultural and tourism strategies. However, as there were no pre-existing strategies, they could not be taken into account. Regional Cultural Consortia were established as coordinating bodies, so they had no real resources to deploy. In the case of South East England the result was *The Cultural Cornerstone* (SECCI, 2001), a very short, general statement of cultural objectives. However, the growing power and influence of the Regional Development Agencies effectively made them the power brokers. Not surprisingly, Regional Cultural Consortia set about making their case relevant to the needs of the Regional Development Agencies; one means of doing this was to promote the cultural industries agenda.

As already noted, the cultural industries agenda had been projected into the limelight with the publication of the *Mapping Document*; it had been given a further boost in a government report on business clusters (DTI, 2000) which highlighted the role that cultural industry clusters may play in regional growth. The problem at the regional level for all parties concerned was that there was no regional data on the cultural industries. Thus, without data policy development and inclusion in the strategic policy process, a cultural industries agenda could not develop. The following section discusses the practical steps taken to create a relevant evidence base.

Constructing an evidence base

The evolution of the new policy environment outlined above created a need for new data on the cultural industries; in turn, this required ever more ingenious ways of extracting relevant data from multiple sources, none of which was really appropriate for the task. Whilst the *Mapping Document* had created a new respect for cultural statistics, the new regional data, which in many ways was more difficult to manipulate, threatened to undermine the legitimacy of the exercise. The regional agencies concerned readily recognized this problem and resorted to contracting the work out to a number of private consultancies. As each of the development agencies created their own regional mapping documents separately, comparability also became an issue. In recognition of this fact, and the obvious need to co-ordinate the exercise, a further group of consultants was commissioned to develop a *Regional Cultural Data Framework* (DCMS, 2003) in order to create a template for all future regional cultural data collection.

Methods, concepts and definitions

Early attempts to measure the contribution of the cultural industries to economies deployed indirect, impact or multiplier analyses (see Pratt, 2001). This approach seemed to support the notion that it was problematic to directly measure the value of cultural activities. The U.K. *Mapping Documents* draw upon a different tradition, one that seeks to measure direct effects. Researchers seeking to measure employment in the arts sector developed some early approaches. O'Brien and Feist's (1995) work, for example, used a combination of occupational and industrial taxonomies to pinpoint only those cultural workers who actually worked in cultural industries; such an approach seeks to highlight only cultural occupations and to ignore the institutional framework within which such work takes place. More recently systemic models of the cultural industries have been advocated that seek to capture the whole "production chain" from inception to consumption. This holistic view seeks to recognize that cultural industries, as do all industries, rely upon, sustain and promote significant manufacturing, distribution and consumption activities. In practice the key difference between the "occupational" and the "production chain" models is that the latter include the range of allied and support activities that make cultural industry outputs possible.

Pratt's (1997; Pratt and Naylor, 2003) model of the cultural industries production system (CIPS) is an example of a production chain model that entails the production system within which cultural goods are produced. It effectively deepens the definition of the creative industries. In the initial formulation four "moments" in the cultural production system were identified:

1. *Content origination.* The generation of new ideas – usually authors, designers or composers – and the value derived from intellectual property rights.
2. *Manufacturing inputs.* Ideas must be turned into products and prototypes using tools and materials, for example the initial recording of a song or the manuscript of a book; these activities necessarily include the production

and supply of things as diverse as musical instruments, film or audio equipment and paint.

3 *Reproduction.* Most cultural industry products need to be mass produced; examples include printing, music, broadcasting and mass production of original designs.
4 *Exchange.* The relationship to the audience or market place. This takes place through physical and virtual retail, via wholesalers and distributors, as well as in theatres, museums, libraries, galleries, historic buildings, sports facilities and other venues and locations.

Later work, drawing upon proposals from the United Nations and the European Union (EU, 2000), has suggested an extension of this concept to six moments by adding the following two:

5 *Education and critique* (to cover both training and the discourse in critical ideas), and
6 *Archiving* (to include libraries and the "memory" of cultural forms). Whilst this 6-phase model is conceptually more robust, data availability constraints mean that it is impractical to use at the present time; however, it serves as an aspirational model of data collection.

The systemic model of cultural production has many similarities with models of innovation such as those described by Lundvall (1992) in that they offer a perspective on the embedded nature of cultural industrial production (see Jeffcutt and Pratt, 2002), and point to the complex web of networks surrounding cultural production (see Grabher, 2003). From a public policy perspective there is additional value in using a systemic model of the cultural industries in that potential points of policy intervention can be reviewed and assessed. The richer understanding of the production process offered opens up the possibilities of identifying strengths, weaknesses, opportunities and threats, as well as an assessment of the sustainability of those activities.

However, as indicated above, the operationalization of a systemic model of cultural production has a number of problems associated with it. The central issue concerns the dominant taxonomies of industries, the Standard Industrial Classification; this classification is used as a basis for all government data collection related to businesses, whether it is employment or output data.[4] In the U.K. the Standard Industrial Classification has a mixed logic: it is partially based upon a final product classification (mostly for manufacturing), and partially based upon an activity classification (mostly for services). Moreover, services are generally described in far less detail, and have fewer unique classification categories, than manufacturing, even though at the current time most economic activity and employment is concentrated in them.[5]

It is an unfortunate fact of life that contemporary industrial classifications are founded upon historic industrial structures; as a consequence the service sector, and the cultural sector, are poorly served, tending to be little more than a residual to manufacturing within such a framework (see Gershuny and Miles, 1983; Walker, 1985). Such a taxonomy has both a powerful rhetorical effect of making these service-sector activities appear to be less significant than they actually are, as well as framing any data produced on the basis of them as poor and imprecise compared to manufacturing figures. This further undermines claims to legitimacy for these industries.

Moreover, there are two categories of industry that are overlooked altogether by the Standard Industrial Classification. First, we can consider the case of new industries. The latest modification of the classification in use in the U.K. is that of 1992; however, the growth of new media, for instance, can be dated from 1993.[6] Second, specialist industries; a good example is the high-fashion industry which the U.K. government considers to be a cultural industry. In the absence of a classification code for "high" as opposed to "high street" fashion it is almost impossible to differentiate this important activity. It is for these, and other similar, reasons that many innovative industries are overlooked in such classifications and measures.

It is for the above reasons that the researcher of the cultural industries has to work with pragmatism and ingenuity. A common strategy has been to carefully comb the industrial classification for activities (4-digit codes) that are wholly cultural and to re-combine them as "the cultural sector." This is the method that was adopted for the analysis presented in this chapter. Annex 1 shows the detailed activities selected, which are then re-grouped along the lines of Pratt's (1997) cultural industries production system into their four functions (as noted above).

The South East example

This section outlines what a regional portrait of the cultural industries looks like, and offers some particular comments on the problems and issues that arise in constructing such a picture. The South East of England, along with all other English regions, was required to produce a regional strategy. The *South East Revised Economic Strategy* was produced in 2002 with a 10-year time horizon (SEEDA, 2002). In order to link with this vision the Regional Cultural Consortia commissioned a report to supplement its broad strategic view: an analysis of the impact of the creative industries in the region (DPA, 2002). This analysis, part of which is reported on here, was based upon a desktop survey of reliable and comprehensive secondary data sources[7] and supplemented by a number of strategic interviews. In this case an expanded definition of what was termed the "cultural industries sector" was used (see Annex 1). The main difference from previous classifications is that aspects of sport and tourism are included in the definition.

Spatial units

The spatial unit that the South East Regional Development has responsibility for does not represent a logical division of social, political or economic space. Essentially, it forms a cordon that runs 270 degrees around London, but it does not include London (see Figure 2.1). As has been extensively discussed elsewhere, there is a very strong case for conceiving of London and the "Rest of the South East" (that is the South East planning region, and the counties of Hertfordshire and Essex which are part of the Eastern planning region) as one functional region (Simmie, 1994). The logical case is that the whole of the South East corner of England functions as a travel to work area and an immediate economic hinterland for London. Accordingly, any analysis of activities that splits London from its region is likely to be problematic. However, the administrative and political reality is a fragmented economic space. The South East region has to both produce its plans, and look to itself as a region. The case for strategic co-ordination is further undermined due to a further disjuncture, namely that London did not produce its regional strategy until 2003, so its plans could not be taken account of. The time lag of three years between the formation of the two agencies and the publication of their strategies[8] was due to the election of a Mayor for London; only when this was achieved could the Greater London Authority and its development agency, the London Development Agency, begin their work.

The relationships between London and the South East region might be expected to be dependent in character, based on the assumption that London "draws in" activities, notably labor and audiences. One might have the further

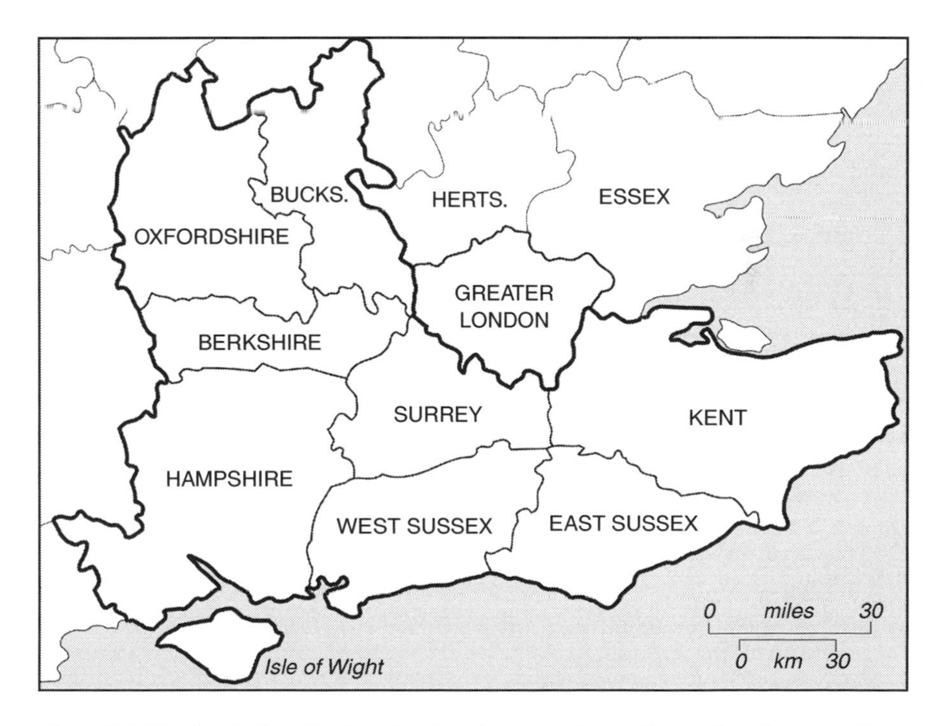

Figure 2.1 The South East Region showing the proximity to Greater London, and Hertfordshire and Essex (both part of the Eastern Region) (source: LSE Design Unit.)

expectation that the South East might be under-represented in the creative industries as a result of this "neighbor" effect. Taking a step further, if the cultural industries followed an industrial logic one might expect the consequences of industrial restructuring in those industries to demonstrate "sorting" of Content Origination activities to London, and Manufacturing Inputs to the outer South East region (where land prices and labor might be marginally lower). It is these issues, in addition to the general nature and dynamics of the cultural industries in the South East, which are explored in the analysis that follows.

An overview of the cultural industries sector

The headline figure that emerged from our analysis was that the South East's cultural industries employed 558,643 people in 2000. This figure positions it as one of the major components of the economy. As Table 2.1 shows, it is only exceeded by employment in Distribution, Hotels and Restaurants; Banking and Finance; and Public Administration, Health and Education. Table 2.1 reinforces the point made above, that if the cultural sector were more statistically visible it might gain greater analytic attention than it has gained hereto. Table 2.2 shows that whilst the number of employees in the cultural sector in the South East is substantial, it is still dwarfed by that of its neighbor London (705,779). Overall, the South East employed 19 percent of the total cultural sector workforce in England. (If the South East and London employment totals are added together they account for 44 percent of the national cultural industries workforce.) We might naturally expect a concentration of all economic activities in London and the South East, the U.K. is notoriously imbalanced in its economic activity. However, just 17 percent of all workers in England benefit from employment in the South East, showing that cultural activities are more concentrated. This point is further strengthened when we note that whilst 13 percent of all employees in England worked in the cultural sector, the figure for the South

Table 2.1 Employment in the major industrial classes in London and South East England, 2000

Industrial class	*London*	*South East*
1: Agriculture and fishing	4,622	40,688
2: Energy and water	13,915	19,163
3: Manufacturing	285,840	432,596
4: Construction	130,584	156,292
5: Distribution, hotels and restaurants	887,840	944,068
6: Transport and communications	317,924	242,630
7: Banking, finance and insurance, etc.	1,360,242	836,251
8: Public administration, education & health	798,585	810,846
9: Other services	261,110	180,951

Source: Annual Business Enquiry (ONS 2003) Crown Copyright

Table 2.2 Employment in the cultural industries (divided by cultural industries production system function), London, South East and England, 2000

CIPS	*London*	*South East*	*England*
Content origination	397,550	256,165	1,191,557
Manufacturing inputs	16,569	37,961	178,301
Reproductive activities	41,290	28,292	188,053
Exchange	161,489	158,806	900,039
All CIPS	705,779	558,643	2,870,345

Source: Annual Business Enquiry (ONS 2003) Crown Copyright

East comparing cultural employment with all employment was higher: 15 percent (17 percent in London).

In common with the national picture, cultural businesses in the South East are characterized by a significant number of very small businesses and a handful of very large enterprises. The precise scale of these businesses is captured by the fact that 86 percent of value added tax registered businesses and organizations turn over less than £500,000 per annum; it is quite likely that many other businesses fall beneath the value added tax threshold (£56,000) and are not represented in these statistics.[9] Correspondingly, the average number of employees per enterprise across the region is about 8 persons, and the average turnover of cultural businesses is approximately £900,000.

The cultural industries are traditionally viewed as industries that have a high proportion of employees in casual, irregular or self-employment. This pattern contrasts with most other industries where such an employment pattern marks participants as marginal; in the cultural sector such patterns can signify core workers (see, for example, Blair, 2001, on the film industry). Compared with the U.K. as a whole where 22 percent of the cultural workforce is self-employed, in London self-employment is 34 percent. The South East has lower than the national average in self-employment, 19 percent of the total cultural workforce. It is not possible to isolate details of freelancing and short-term contracts from the sources available. At the other end of the spectrum from freelance and (serial) short-term contract workers are part-time workers. These workers do tend to be more marginal: 30 percent of the cultural workforce in the South East are part-timers, which is the second lowest proportion of all English regions (behind London: 21 percent).

In summary, the picture is of a large number of people involved in an industry that is dominated by micro-enterprises with a small turnover. Furthermore, that labor market has a large proportion of self-employed (freelance and contract) workers in it. Cultural businesses certainly might be considered as glamorous activities that attract high wages, and the existence of a few superstars could skew earnings data significantly. On the other hand, given the domination of the sector by small enterprises and freelancers, it is perhaps surprising to find that, on average, full-time weekly wages in the sector are 20 percent higher than the regional average for the whole workforce. Given general English regional disparities, especially in living costs, it is perhaps less surprising to learn that full-time cultural workers in the South East earn 18 percent more than in other regions.[10]

Perhaps another surprising element of the cultural sector is educational attainment. The received image might be, at best, one of craft workers with little formal education. Moreover, in general terms we normally expect that a small-firm, self-employed and freelance-dominated workforce would not register high levels of educational attainment. In fact, the real picture is quite the opposite, one of the characteristics of the cultural sector is the high level of qualifications held by the workforce, albeit not always the "relevant" qualification for their current job. Thirty percent of those working in the cultural sector hold a degree or equivalent, compared with 20 percent in the South East economy more generally.

Beyond these characteristics one vector stands out amongst all others: employment growth. Table 2.3 shows that in the period 1995–2000 employment growth in the cultural sector was 22 percent in England but a substantial 34 percent in the South East, a figure which eclipses even that of London at 29 percent. These data contrast with national figures on all employment growth for the period of 16 percent, and of 24 percent and 23 percent in the South East and London respectively: London and the South East grew at a significantly faster rate than all other regions.

Whilst one might not have begun an analysis by looking at the South East separately from London, the findings reported show that several unexpected things seem to be happening. The first and most obvious point is of both the size and growth of the cultural sector that is hinted at in the *Mapping Document*. Here we can note that not only is the growth impressive, but that it also outstrips that of much of the rest of the economy too. We might have expected London to have a major concentration of cultural sector activities; however, the data points to even greater concentrations occurring than in other sectors. It is clear that there is a strong regional-level dynamic operating in the cultural sector. When one contrasts the experience of the South East and London there

Table 2.3 Employment change (percent) in the cultural industries (divided by cultural industries production system function), London, South East and England, 1995–2000

CIPS	*London*	*South East*	*England*
Content origination	33.8	39.9	26.6
Manufacturing inputs	18.1	13.4	6.8
Reproductive activities	1.9	−3.4	−3.3
Exchange	9.0	24.9	12.8
All CIPS	28.9	33.7	22.3

Sources: Annual Business Enquiry; Census of Employment (ONS 2003) Crown Copyright

are some less expected findings. Cultural industries in the South East have been growing at a faster rate than those in London. In order to tease out some potential explanations for this differential growth I turn next to an analysis based upon the functions of the cultural production system.

The cultural industries production system perspective

By looking at the separate functions of cultural production a number of dynamics are pointed up. The first point, as can be seen in Table 2.2, is that Content Origination accounts for a major part of employment in the sector, followed by Exchange activities. The national average in 2000 was 41 percent employed in Content Origination activities; 45 percent of cultural workers in the South East and 56 percent of those in London are employed in this function. Moreover, as might be expected, the proportion of Manufacturing Inputs is higher in the South East (7 percent) compared with London (2 percent).

A second point concerns the dynamics of change. Table 2.3 highlights the fact that much of the positive growth in the cultural sector at the national level can be accounted for by Content Origination. The South East replicates this pattern, but with exaggerated growth in all areas except Reproduction activities, where the decline is in line with national trends. London contrasts with the South East only in the fact that it registers positive growth in Reproduction Activities but a relatively low growth in Exchange functions.

Thus, within this pattern of growth, the South East has been forging ahead; far from being a laggard, or simply a "back region" of support activities, it does seem as though dynamism in Content Origination activities is much in evidence. Reproduction Activities, the closest function to mass production in the cultural industries, are declining faster in the South East than in London. This raises another question, one that we cannot explore here, namely that "supporting functions" may be driven to more peripheral regions, or abroad entirely. This issue would repay further investigation as it potentially throws into doubt the notion that there may be strong inter-dependencies within the whole production system.

It is worth noting that the other positive dynamic in cultural employment are Manufacturing Inputs – these activities show positive growth nationally, and they are far stronger in London and the South East. Given the massive across-the-board declines in manufacturing industry, especially in the South East, this growth must be one of the few sectors of manufacturing experiencing growth. This is indicative evidence of the interrelated nature of Manufacture and Content Origination activities across the production system.[11]

Finally, we can illuminate some aspects of the value chain. Generally, such data is very difficult to collect. The best we can do here is to look at turnover data; ideally one would examine input–output data. Examining the turnover of firms (which is £47bn for the cultural sector in the South East), and breaking them down by function, we can find supporting evidence for the employment findings reported above; namely that the value of cultural sector Manufacturing functions are robust, accounting for 36 percent of turnover, as is that of Content Origination which accounts for 37 percent of the turnover. This underlines the fact that cultural-sector-related Manufacturing activities, as measured here, are still important even in the cultural economy and it is allied to solid growth in Content Origination. Moreover, despite the fact that most of the Content Origination companies are very small, compared to the medium-sized Manufacturers, they still produce a significant contribution to turnover for the sector.

Clusters

Some intra-regional dimensions of the cultural sector are considered in this section. An obvious and striking point is that employment in Exchange functions tends to be focused on the major areas of population. Additionally, and echoing broad patterns of economic activity in the South East, there was a focus of manufacturing activities in the East of the region. The South East's traditional engine of growth has been focused along the western corridor of the M4 motorway (see Hall *et al.*, 1987) and linked to the defense industry and technology-orientated firms. In general terms the emergent focus of the cultural industries does not diverge significantly from the broad north-western sector of existing growth in the region.

All of the foregoing raises questions about the micro-spatial dimensions of cultural industries activities: clustering. One of the key characteristics of the cultural industries is the proclivity of firms and labor to cluster together (see Scott, 2000). The secondary data sources that this chapter is based upon are all but exhausted prior to such a detailed level of analysis. Accordingly, it was supplemented with indicative interview and documentary research. This investigation highlights the existence and general character of a number of clusters of the cultural sector in the region. However, it is clearly something that can only be properly investigated through more intensive research. A well-documented example of a regional cluster is that of "motor sport valley" associated with Formula One motor-racing teams (see Henry *et al.*, 1996; Pinch and Henry, 1999); a cluster of computer games companies exists immediately south of London in Guildford (see Human Capital, 2001); there are new media companies on the south coast in Brighton (see DPA, 2001; Pratt and Gill, 2000); and there is a significant publishing cluster based around Oxford in the north-west of the region. Arguably, "motor sport valley" has a close relation in the Recreational Marine Industries, mainly located in Hampshire. Like Formula One, Recreational Marine Industries have their sporting events, for example Cowes Week, which helps to sustain a demand for state of the art materials and technologies. These technologies overlap with Formula One, and those of the marine defense industries also located in the region; finally, the industry also has a popular recreational activity linked to sailing and marinas that are scattered along the south coast. Unfortunately, all of these industries, some more familiar than others, are all but overlooked by statistical surveys; where they are picked up it is only in a partial manner. These clusters of significant cultural activities were not identified by secondary data sources on the whole, the reason being that the industrial classification taxonomy is insensitive to the complex industrial filieres of the production system that they represent. Additionally, quite simply, there is no industrial classification code for Formula One racing, computer games or recreational marine industries. Given the generally robust nature of the cultural sector and the outstanding performance of particular industries associated with clusters, there is clearly value in further investigation.

The evidence presented earlier in this chapter shows elements of a set of highly robust cultural industries existing in the South East region. Significant growth can be noted in those fragments of industry that can be measured; coupled with anecdotal evidence it is suggestive of strength in still other elements. It is striking that strong performance can be identified in both the Content Origination and Manufacturing activities, suggesting, as the conceptual model would lead us to expect, that there might be a significant positive relationship between them. Contrary to some superficial readings of post-industrial society, the knowledge economy (or, in this case, a specific fragment of it, the cultural industries) has not severed its links with manufacture. On the basis of the findings reported here there is a significant need for research that can explore the precise social and economic character of spatial co-dependencies within and between these industries that will disclose the nature of the tendency to clustering and what advantages it confers.

Conclusion: the art of the possible

The U.K. government has garnered a number of headlines around the world by initiating a more systematic analysis of its cultural industries. Initial findings projected the cultural industries as playing a significant role in the economy, and contributing to export earnings. Later, indications of substantial growth further fuelled expectations of the sector. However, at the same time these headline figures pointed to the lacunae of detailed analysis and understanding of processes that may reliably explain the nature of change. These debates, and the initial "mapping" that has gone along with them, have been picked up in a number of nation states. It is for both reasons – the popularity of the approach and the attractive economic growth that it signals – that the next stage of the U.K.'s policy development may be of wider interest.

This chapter outlined how the internal dynamics of the U.K. political system had led to the development of a regionalization program across all policy areas, the cultural industries included. The major outcome, in the U.K. context, of these pressures has been the coupling of the cultural industries to a regional economic development agenda. A subsidiary outcome has been the development of regional data collection and analysis, and the creation of a more robust methodological framework for data collection on the cultural industries. The chapter discussed the findings of an analysis of change in the cultural industries in South East England between 1995–2000.

It will come as no surprise to anyone that the standard administrative region of the South East is not a particularly useful basis upon which to analyze the cultural industries. However, the chapter used this template, as policy must be developed for this spatial and administrative unit. It was expected that the findings would point to the South East region playing a subservient role to London in the case of the cultural industries. It was initially assumed that more extensive land users and more labor-intensive activities might be found in the South East, and high-prestige content production concentrated in London.

The findings indicated that the South East was growing faster than London as regards the cultural industries; Content Origination activities were found well represented in outer South East as well as inner London areas. There did seem to be some significant evidence that these activities were tightly clustered, but that cultural industry clusters were to be found in many locations across the South East and in London. The obvious question is: why has this growth occurred? Certainly, the U.K. has experienced a period of rapid employment growth; however, the cultural industries have bettered this, in the South East and London in particular. There is no simple answer as to why this has happened. It is not, as yet, directly policy related as there has been little, if any, cultural industries policy implemented in the period under study. Clearly, more detailed, and longitudinal, analysis will be required to begin to answer this question. One point that did emerge from the analysis that may inform future analysis: there did seem to be a co-incidence of growth in manufacturing, exchange and content origination activities. This points to the possible value of closely linked production functions across the South East and London; these activities appear to be rooted in the many local nodes or clusters. At the regional scale, policy making for the cultural industries will most likely develop as an adjunct of economic policy. The challenge ahead for policy makers is two-fold: will the cultural industries respond to standard industrial policies, or will specific policy be required?; and how will regions co-ordinate their policy-making activities with one another in the field of the cultural industries?

However, given the failure to account for the dramatic growth that has occurred in the sector in recent years, some more basic analysis would seem to be called for. The small steps that the U.K. has taken thus far may be described as the promotion of a "way of looking" at the cultural sector. In many respects a very subtle type of policy is forming, one of making visible the activities of the cultural sector; once made visible these activities may be made "governable". It is important to note that whilst a whole section of the economy has been made visible, the "non-economic" aspect of that economy has yet to see the light. The cultural industries rely, in practice, on rich webs of traded and un-traded relationships. We have as yet scratched the surface of registering some of these traded, and trade-related, relations. The un-traded relationships remain, sadly, invisible to our analyses. The challenge for the future is a systematic use of more qualitative techniques and primary modes of investigation to explore these un-traded activities. In the cultural industries analyses, un-traded activities may take us back into an understanding of not-for-profit and "fine arts", and an exploration of their articulation to the more commercial activities that the cultural industries commonly refer to.

The U.K.'s shifting policy environment and its increasingly regional emphasis has provided an incentive to explore a smaller-scale of analysis of the cultural industries. The national level analysis has generated considerable interest and the findings here suggest that regional scale analyses will throw up more useful questions. However, the biggest challenge facing researchers and policy makers is to put in place more adequate information collection protocols. The initial attempts at "re-purposing" secondary sources have been revealing and suggestive, but they are close to exhaustion; now more comprehensive data is required. Additionally, more detailed information is required, information that engages with subtlety of interaction, communication and proximity across the whole production chain. This latter form of data will be the most resource intensive to collect; however, at present it represents the most pressing need if a useful and informative evidence base is to be constructed.

Annex 1: the cultural industries sector

Below is a list of industrial codes of the cultural industries used in the analysis and their classification using the cultural industries production system. All 4-digit industry codes refer to the Standard Industrial Classification (1992).

Content origination

2211: Publishing of books
2212: Publishing of newspapers
2213: Publishing of journals and periodicals
2214: Publishing of sound recordings
2215: Other publishing
7220: Software consultancy and supply
7420: Architectural/engineering activities
7440: Advertising
7481: Photographic activities
7484: Other business activities
9211: Motion picture and video production
9220: Radio and television activities
9231: Artistic and literary creation, etc.
9240: News-agency activities

Manufacturing inputs

2464: Manufacture of photographic chemicals
2465: Manufacture of prepared unrecorded media
3210: Manufacture of electronic valves, etc.
3220: Manufacture of TV/radio transmitters, etc.

3230: Manufacture of TV/radio receivers, etc.
3340: Manufacture of optical instruments, etc.
3512: Building repairing of pleasure boats, etc.
3622: Manufacture of jewelry
3630: Manufacture of musical instruments
3640: Manufacture of sports goods
3650: Manufacture of games and toys

Reproduction

2221: Printing of newspapers
2222: Printing
2223: Bookbinding and finishing
2224: Composition and plate-making
2225: Other activities related to printing
2231: Reproduction of sound recording
2232: Reproduction of video recording
2233: Reproduction of computer media
9212: Motion picture and video distribution

Exchange

5143: Wholesale of electrical household goods
5245: Retail sale: electrical household goods
5247: Retail sale of books/newspapers, etc.
5511: Hotels and motels, with restaurant
5512: Hotels and motels, without restaurant
5521: Youth hostels and mountain refuges
5522: Camping sites, including caravan sites
5523: Other provision of lodgings
9213: Motion-picture projection
9232: Operation of arts facilities
9233: Fair and amusement-park activities
9234: Other entertainment activities
9251: Library and archives activities
9252: Museum activities, etc.
9253: Botanical and zoological gardens, etc.
9261: Operation of sports arenas and stadiums
9262: Other sporting activities
9271: Gambling and betting activities
9272: Other recreational activities

Notes

1 The term cultural industry is used in this chapter. The U.K. government has preferred the use of the term "creative industry." Whilst there are some significant differences in the usage and meaning of the terms this need not concern us here.

2 Australia, New Zealand and Canada have also collected statistics on the cultural industries, however, they have not stressed the economic role so clearly as the U.K.
3 Initially the Regional Development Agencies were given a guiding role only. From 2002 onwards they have also been given substantial funding and the power to allocate those funds according to their own agenda.
4 Occupation is classified via a separate taxonomy, the Standard Occupational Classification; the problems of this echo those of the industrial classification discussed here.
5 These issues are discussed in more detail in Pratt (1997, 2000)
6 It can take at least 15 years for a "new industry" to gain some visibility in the Standard Industrial Classification. Even the revision of the classification due in 2007, which was under consultation in 2003, will still under-represent the cultural industries.
7 The sources used are the Annual Business Inquiry, Annual Employment Survey, Labour Force Survey, Inter-Departmental Business Register, and were supplemented by data from both Yellow Pages Business Data and Companies House Registers.
8 See GLA (2003), and GLA Economics (2002)
9 All data on turnover presented here is derived from DPA (2002).
10 This must be attenuated by the fact that the South East has a higher cost of living than the rest of the U.K., this figure is about average.
11 Clearly, this is a co-relation; evidence of causality would require further complementary qualitative research

Bibliography

Blair, H. (2001) "'You're only as good as your last job': the labour process and labour market in the British Film Industry," *Work Employment and Society*, 15: 1–21

DCMS (1998) *The Creative Industries Mapping Document*, London: Department of Culture, Media and Sport.

—— (2001) *The Creative Industries Mapping Document*, London: Department of Culture, Media and Sport.

—— (2003) *Regional Cultural Data Framework*, a report for the Department of Culture, Media and Sport prepared by Positive Solutions, Business Strategies, Burns Owens Partnership and Andy C. Pratt, London: Department of Culture, Media and Sport.

DPA (2001) *Brighton and Hove Creative Industries Report*, a report for Brighton and Hove Council, London: David Powell and Associates.

—— (2002) *Creative and Cultural Industries: An Economic Impact Study*, a report by David Powell and Associates for South East England for South East Cultural Consortium and South East England Development Agency, Guildford: SEEDA.

DTI (2000) *Business Clusters in the U.K.: A First Assessment*, London: Department of Trade and Industry.

European Union (2000) *Cultural Statistics in the EU: Final Report of the LEG*, Brussels: European Union.

Gershuny, J. and Miles, I. (1983) *The New Service Economy: The Transformation of Employment in Industrial Societies*, London: Pinter.

Grabher, G. (2003) "Cool projects, boring institutions: temporary collaboration in social context," *Regional Studies*, 36: 205–14.

Greater London Authority (2003) *London Cultural Capital: Realising the Potential of a World-class City*, London: Greater London Authority.

Greater London Authority Economics (2002) *Creativity: London's Core Business*, London: Greater London Authority.

Hall, P., Berheny, M., McQuaid, R. and Hart, D. (1987) *Western Sunrise: The Genesis and Growth of Britain's Hi-tech Corridor*, London: Allen and Unwin.

Henry, N., Pinch, S. and Russell, S. (1996) "In pole position? Untraded interdependencies, new industrial spaces and the British motor sport industry," *Area,* 28: 25–36.

Human Capital (2001) *The U.K. Games Industry and Higher Education*, a report for DTI, London: Human Capital.

Jeffcutt, P. and Pratt, A. C. (2002) "Managing Creativity in the cultural industries," *Creativity and Innovation Management*, 11: 225–33.

Lundvall, B. (1992) *National Systems of Innovation*, London: Pinter.

O'Brien, J. and Feist, A. (1995) *Employment in the Arts and Cultural Industries: An Analysis of the 1991 Census*, ACE research report no. 2, London: Arts Council of England.

Pinch, S. and Henry, N. (1999) "Paul Krugman's geographical economics, industrial clustering and the British motor sport industry," *Regional Studies,* 33: 815–27.

Pratt, A. C. (1997) "The cultural industries production system: a case study of employment change in Britain, 1984–91," *Environment and Planning A*, 29: 1953–74.

—— (2000) "Employment: the difficulties of classification, the logic of grouping industrial activities comprising the sector, and some summaries of the size and distribution of employment in the creative industries sector in Great Britain 1981–96", in S. Roodhouse (ed.) *The New Cultural Map: A Research Agenda for the 21st Century*, Leeds: Bretton Hall, Leeds University

—— (2001) "Understanding the cultural industries: is more less?," *Culturelink,* Special issue: 51–68.

Pratt, A. C. and Gill, R. (2000) *New Media User Networks*, a report for the Arts Council of England, London: Arts Council of England.

Pratt, A. C. and Naylor, R. (2003) "Winning customers and business: improving links in creative production chains," evidence given to the Mayor's Commission on Creative Industries, available online at: <http://www.creativelondon.org.uk/commission/evidence/index.htm> (accessed 15 October 2003).

SECCI (2001) *The Cultural Cornerstone*, Guildford: SECCI.

SEEDA (2002) *Revised Economic Strategy*, Guildford: SECCI.

Simmie, J. (1994) "Planning and London," in J. Simmie (ed.) *Planning London*, London: UCL Press.

Scott, A, (2000) *The Cultural Economy of Cities*, London: Sage.

Siwek, S. (2002) *Copyright Industries in the U.S. Economy: The 2002 Report*, Washington: The International Intellectual Property Alliance.

Tomaney, J. (1999) "New Labour and the English Question," *The Political Quarterly*, 70: 75–82.

—— (2000) "Debates and developments – end of the empire state? New Labour and devolution in the United Kingdom," *International Journal of Urban and Regional Research,* 24: 675–88.

Walker, R. (1985) "Is there a service economy?," *Science and Society,* 49: 42–83.

3 Cities, culture and "transitional economies"

Developing cultural industries in St. Petersburg

Justin O'Connor

This chapter looks at some issues around the transfer of cultural industry policy between two very different national contexts, the U.K. and Russia. Specifically it draws on a partnership project between Manchester and St. Petersburg[1] financed by the European Union as part of a program to promote economic development through knowledge transfer between Europe and the countries of the former Soviet Union. This specific project attempted to place the cultural industries squarely within the dimension of economic development, and drew on the expertise of Manchester's Creative Industries Development Service (CIDS) and other partners to effect this policy transfer.[2]

The project has been about introducing and inserting a new discourse of "cultural industries" into St. Petersburg (and thus Russian) cultural policy – and attempting thus to move the field of cultural policy to a more central position within the local policy field. This implies a real shift in thinking amongst policy officials and cultural institutions, and equally – given the focus on small cultural producers – the self-understanding and identity of the cultural sector as a whole. I will discuss some of the specific difficulties of cultural industry development initiatives later in the chapter, but I first want to reflect on the processes involved in such policy translation or transposition. This will highlight some wider ambiguities and tensions involved in the discourse around cultural industries with which this book is concerned. Briefly we can say that whilst those who promote the notion of cultural industries mobilize clear economic development objectives, these do not necessarily exhaust all that is involved in this discourse of "cultural industries" – either for those promoting cultural industry policies or for the "sector" who are nominally the objects of such policies. This can be seen clearly in the Russian context which, for reasons we will make clear, necessitates a much more explicit outlining of fundamental principles in order to establish the basic common understanding underpinning the transfer of "technical know how" and "best practice."

Cultural industries as a discursive construction

The "cultural industries" are a discursive construction. The term itself has a complex history, and has constantly demanded clarification as to what it actually means. At one level this involves a definition of "what's in" and "what's out." This is the core of the statistical debate which is intended to provide more quantitative knowledge of the sector as a basis for intervention although it is just as often used by its proponents to give economic weight to a sector seen as economically marginal or volatile. Whilst the details of the statistical debates need not concern us here (O'Connor, 1999a), underlying these is a deeper debate about how we construct "the economic". Beneath the Standard Industrial Classification and Standard Occupational Classification codes lies a shift from a nineteenth-century model of extraction/production of raw materials; manufacturing, transformation, distribution and consumption to a more fluid and complex model in which information processing and volatile consumer perceptions impact at all points of the production chain. This is linked to wider debates about the boundaries of "culture" and "economics" and claims as to their mutual transformation which cannot be broached here (du Gay and Pryke, 2002). Defining the cultural industries is at once an exercise in constructing the boundaries of an emergent sector and in situating this sector on the restless tectonics of the new "economy" and the new "culture". At a policy level this is a ***work*** that needs to be done and in so doing it mobilizes narrative to provide its dynamism and vision. Defining the cultural industries is about telling a convincing story about the meaning of contemporary culture – this does involve a story about a new economics but I would argue it is the new, more "central," role for culture that this implies which is primary.

This is how it was used by the Greater London Council (GLC) in the early 1980s, who attempted to put Adorno's polemical – and deliberately outrageous – term to more positive work (Adorno and Horkheimer, 1944). It was used to outline a vision of a democratic cultural policy (and politics), where financial support for the traditional high arts was challenged by an increased legitimation of popular culture and its forms of production. In an attempt to somehow support the latter, the GLC were forced to go beyond the paradigm of support for the individual artist (whether directly or via institutions) and look at intervention in the "sector" as a whole. And they used the language of industry sectors and economic development, and especially the "value chain" as developed by academics in media and cultural studies, to map this out. The intention was certainly local economic development, but it was also about the support and promotion of the creative (and thus smaller and more local) end of production – giving those involved more power over their products and more economic returns on their labor. These strands were often vague and confused, and certainly under-developed, but they outlined a policy that was not simply about economics – it was a democratic and local (albeit London centered) cultural policy.

A different narrative is mobilized around the current terminological shift in the U.K., and increasingly in Australia, New Zealand, Canada and the USA, where it has now been superseded by "the creative industries." This term was most famously coined by the U.K. government's Department of Culture, Media and Sport (DCMS) in 1998. If a recent scholarly text defined the "core" cultural industries as those that "deal with the industrial production and circulation of texts" (Hesmondhalgh, 2002: 12) and others have pointed to "those industries whose primary economic value is derived from their cultural value" (O'Connor, 1999a: 34), the DCMS used a definition with severely limited analytical power but one that mobilized some key buzz words, situating the cultural industries temptingly on a new economic territory: "those activities which have their origin in individual creativity, skill and talent, and which have the potential for wealth and job creation through the generation and exploitation of intellectual property" (DCMS, 1998: 3). The shift from "cultural" to "creative" set the seal on this, moving from a word that had elitist connotations in the U.K. to one that annexed a whole literature of new management-speak as well as the dynamism of that young, alternative world disaffected by Thatcherism and whose symbol was taken to be "Brit Pop" (Harris, 2003). The term was very much linked to the British Prime Minister Tony Blair's "Third Way" and to the discourse of entrepreneurship which had been promoted by Thatcher's government but was now reworked away from the nostalgia of Victorian "self-made men" to the youthful creativity of the post-1960s generation (Leadbetter and Oakley, 1999). It was part of an attempt to wrest the narrative of modernity and change from the Right which could be seen at its clearest in the "New Times" debates of the late 1980s (Hall and Jacques, 1989).

Policy advocates as cultural intermediaries

Hesmondhalgh's (2002) account of the debates around the cultural industries is based exclusively on academic texts; this ignores the extent to which the term was *made to work* – given meaning and operationalized across different policy terrains and in the service of different interest groups. The primary site for this, I would argue, is the local city – and latterly regional – level. The late 1980s and early 1990s saw a migration of cultural industry arguments from the GLC to other metropolitan regions in the U.K. – through both consultants and policy makers. They became intertwined with local agendas and narratives of decline and rebirth in different ways and with different success. Indeed, despite the increasing globalization of the cultural industries in the last twenty years the story of cultural industry policy is very much one of the local and urban. It could also be argued that this construction of a new policy object, this re-configuring of the cultural policy field, moving it much closer to the "hard hitting" fields of economic development, urban regeneration and (to a lesser extent) social policy has also seen the distinct cultural politics of the GLC policy – its stress on the local and the independent being so important for its subsequent pursuit elsewhere – increasingly buried under the discourse of economic development.

One other key dimension also goes unnoticed in Hesmondhalgh's book, which is that the sector itself, the nominal object of both academic and policy discourse does not recognize itself in the term "cultural industries" – at least not immediately. Some are simply unaware of how their activities relate to a range of disparate occupations and businesses. Some are explicit in their refusal of the terminology and the company with which they are thus grouped. Indeed, one of the key arguments of the policy advocates is that this sector lacks a necessary voice, it needs to express its demands, needs to become self-conscious as a sector, needs to present itself with the coherence of other economic groups, needs, therefore, to co-operate in its own construction as policy object (O'Connor, 1999a). If an essential part of this discursive operation is the dismantling of fixed oppositions between economics and culture then this has to be about the self-perception, identity (and identification) of cultural producers – the inculcation or adoption of a new kind of what Nigel Thrift calls "embodied performative knowledge" but can also be seen as a form of habitus (O'Connor, 1997, 1999a, 2000b).

The cultural industries discourse then is not simply policy making but is part of a wider shift in governance, and requires a new set of self-understandings as part of the key skills in a new cultural economy (O'Connor, 2000b). In this sense those concerned to advocate cultural industry strategies could be seen as a species of "cultural intermediaries." As has recently been pointed out (Hesmondhalgh, 2002; Negus, 2002) this term has been used with some quite divergent meanings or at least divergent from Bourdieu's original intentions (Bourdieu, 1986). If they can be seen as active and conscious agents of social and cultural change – change in the perceived interests of themselves and the class fraction to which they belong, as Bourdieu would have it – then they have been identified with cultural critics (Bourdieu, 1986; Hesmondhalgh, 2002); those promoting a new lifestyle (Featherstone, 1990; O'Connor and Wynne, 1998); and those who chose which products go forward through the cultural production chain (Negus, 2002). It has also been used to describe those who "make things happen," putting artists, money and audiences together in a way that creates new cultural possibilities. This might include Diaghilev, or Brian Epstein, or Charles Saatchi. At a more mundane level it can be used to describe those who are able to translate between the language of policy makers and that of the cultural producers. As with the A+R men (music industry talent scouts: Artists and Repertoire) in Negus' description these intermediaries work to link one level of discourse to another – to "represent" the interests of cultural producers within the context of wider policy development, and speak this language back to those producers.

By the time Tony Blair's New Labour came to power in 1997 in the U.K. the cultural industries had a strong policy presence – it was here that consultants and policy makers had translated academic literature and practical examples into coherent policy possibilities. The narrative context for this was boosted by New Labour's legitimizing of the cultural industries – and the term "creative" allowed an argument about a benign conjunction of culture and economics to be situated at the level of personal potential and aspiration. Those in the sector could now recognize themselves and others as "creatives" (Leadbetter and Oakley, 1999; Caves, 2000; Florida, 2002).

At the same time the cultural industries also became a U.K. policy export, with consultants – and now academics – being invited by many European cities to advise on culture as a motor of economic development. However, the interaction of these policy intermediaries with very different contexts meant that the work of definition had to be done over, and as such the narratives spelled out more clearly. Often this was not easy as the cultural (and by now "creative") industry discourse was associated with Blair's "Third Way," or with some Anglo-U.S. assault on a European cultural policy consensus. Indeed it was quite clear that a shift in discourse would challenge established policy consensus. The terminology itself brought fresh problems; whereas the U.K. can use "industry" almost interchangeably with "economic sector," elsewhere it evokes factory production (O'Connor, 2000). Cultural enterprise or cultural business often had to supplement the main term. In fact "cultural industries" became very much an imported neologism, given in the English original and then explained (O'Connor, 1999b).

How the term and the arguments are used and reconfigured depends on the local context, and we cannot go into detail here. But if it was generally seen as an argument about a new relationship between culture and economics, how this relationship was understood could be very different, as could too the outcomes envisaged and the groups who picked up the ideas. Policy makers used it to drive different agendas – job creation, urban regeneration, the commercialization of subsidized culture, developing new media industries, creating employment, retaining talent, etc. But cultural producers also reacted in different ways – some seeing it as a new set of opportunities, others as the thin end of a dangerous wedge.

It should be clear then that in working to construct a new policy object, and in attempting to shift discourses around culture towards economics – with the proviso that economics too is moving towards culture – the cultural industries discourse mobilizes a narrative to underpin its policy goals. These narratives become more apparent when the discourse enters a new context – it has to justify itself and make its arguments clear not simply as technical policy tools but as concerned with the fundamental direction and meaning of contemporary culture.

St. Petersburg: city of culture

St. Petersburg is Russia's second city, with a population of nearly five million. Founded in 1703 by Peter the Great as the capital of a new, modernizing Russia, it acts as a powerful and complex symbol of the country's relationship to the West. Moscow, the medieval capital which regained this status in 1918 during the revolutionary civil war, is by far the biggest city in Russia, with over 10 million people. It is the economic powerhouse – almost an economic region unto itself – as well as being the center of political power. It is also the largest centre for the big cultural industries – broadcasting, newspapers and magazines, film, music recording, fashion and design. However, St. Petersburg retains its symbolic role as the city of culture. This relates to its undisputed pre-eminence in the field of classical culture, with over 3,000 historic buildings in its central area, and a range of world famous institutions including the Hermitage, the Maryinsky Theatre and the Russian State Museum.

This classical heritage is one dimension of the Petersburg "mythos'; that of "victim city" is another (Volkov, 1995). The shift of power to Moscow, and the devastation of the city in the revolutionary civil war (1918–21) were often portrayed as a form of punishment. The notion of imperial hubris visiting disaster on its inhabitants was embedded in the Petersburg mythos by Pushkin's famous poem of *The Bronze Horseman* which evokes the suffering inflicted by the building of the city and, by implication, the forced modernization of the country. Added to this was a sense of the city paying for its decadent sins in the poetry of Akhmatova, and indeed in the pronouncements of the Communist leaders in Moscow who associated the city with Tsarism and dangerous Western ideas. The "Great Purge" of 1937 began in the city and focused on its intelligentsia; the catastrophe of the 900-day siege by the Nazis did not stop subsequent purges and attacks on the city from Stalin (Berman, 1984; Clark, 1995; Volkov, 1995; Figes, 2002).

The "victim city" thus had long historical roots, although there was also a more mundane "second city" syndrome involving – in common with many other cities proud of a distinct heritage – a resentment at the dominance and perceived self-aggrandizement of a larger, more powerful capital city. Moscow – "Mother Moscow" – had always been set against the willful, artificial, un-Russian Petersburg; Moscow was the spiritual home abandoned by Peter in his pursuit of Western knowledge and values. In the early twentieth-century, St. Petersburg (or Leningrad, as it had become in 1924) became associated with the preservation of the spiritual values of European culture in the face of Moscow's "Asiatic" barbarism. In more recent times this shifted again, with Moscow now being characterized as a city given to the dirty business of money and politics, home to the brash *novye russkie*, the new Russians, spending without taste or culture. St. Petersburg became the city of culture; charged with defending the values of this culture against the plutocracy of the new Moscow, preserving a unique classical heritage in the face of budget cuts and mass culture (and in some iterations representing a "time capsule" or "ark" which European culture itself had long forgotten).[3] A less bombastic version of this saw the city as a place where people could pursue cultural interests for their own sake, having a more laid-back, bohemian lifestyle and attitude, and a more cynical, pessimistic face to present to the shallow optimism of Moscow (Nicolson, 1994; Volkov, 1995).

It would be unwise to underestimate the power of this mythos of St. Petersburg; such narratives show up in the talk of people from well outside the social and cultural elites. However, the 1990s witnessed a severe economic crisis which dominated the political horizon. Like most Russian cities its situation in the 1990s was dire. In particular its main industries, based on military production, collapsed as a result of demobilization after 1991. Unemployment, industrial decline, a massive implosion of self-identity – these were all worse than anything suffered in the rust bowls of Europe and North America. Petersburg had the same infrastructure problems as other Russian cities – roads, housing, transport, water and sewerage, etc. – but it also had a unique collection of historic buildings, over 40 percent of which were in need of urgent repair (Leontief Centre, 1999; Danks, 2001). These buildings had been chronically starved of funding at least since the 1970s, with most of the hard currency earned by *Intourist*, the state-owned travel agency, being retained by central government in Moscow.

The global distinctiveness of these buildings and the cultural heritage they represented was mobilized very effectively in the "re-branding" of the city by the Mayor Anatoly Sobchak, who steered through the renaming of the city in 1991 and embarked on a goodwill tour around various European cities (Causey, 2002). Sobchak attempted to use St. Petersburg's cultural profile to present a modern, dynamic and democratic image for the city, much in the same way as Pasqual Maragall had done for Barcelona at the end of the 1980s. However, Sobchak, under trumped-up accusations of corruption, was voted out of power in 1996, and the city administration became dominated by corrupt and inefficient ex-apparatchik personnel as in many other Russian post-socialist cities (Andrusz *et al.*, 1996; Mellor, 1999). The city became one of the worst places in the country for violent – often gang-related – crime and failed to tackle its basic physical, social or economic infrastructure problems. It was Moscow who got the charismatic Mayor in the form of Yuri Luzhkov.

Some improvements were to be found as the city began to develop new service-sector businesses – which now represents 60 percent of the city's GDP – and tried to bring investment into the city centre (Leontief Centre, 1999). But the recent tercentenary was generally perceived to be a huge wasted opportunity, failing to deliver the improvement in tourist infrastructure (transport, information, affordable hotels, visa restrictions, etc.) or to pass down benefits to the cultural infrastructure outside those of the big global brands. The fact that President Putin was Deputy Mayor to Sobchak, and many of his close advisors are drawn from his former St. Petersburg administration (as well as the city's FSB – the successor to the KGB), has certainly meant that St. Petersburg now feels more investment and attention may come its way. The election of Putin's protégé as new Governor (title changed from Mayor in 1996), with a brief to end corruption and improve efficiency in the administration, is also something that points this way.

Cultural policy in St. Petersburg

The first half of the 1990s saw attempts to place cultural institutions on a more "normal" footing – formalizing their property ownership, removing perks and privileges, detaching them from the leisure and welfare structures of the cultural unions. But from 1996 the main problem facing cultural policy in St. Petersburg was the rapid reduction in state funding for culture, which fell 40 percent between 1991 and 2001. Less than 1 percent of a (until recently) shrinking or static GNP was now spent on culture (Belova *et al.*, 2002). In 1996 and 1998 this had fallen to 0.29 percent and 0.32 percent respectively. Wages were hardly paid, and if so in arrears; buildings and collections were in serious danger. Faced with sheer necessity managers of cultural institutions had to find new sources of funding; if this was sometimes portrayed as market forces unlocking new resources it was mostly felt for what it was – a chaotic scramble (Causey, 2002). In this process the city administration provided barely minimal leadership or strategic framework. Foundations began to provide new sources of funding – the Open Society Institute (Soros Foundation), the Ford, Eurasia and Getty Foundations, and other smaller funders – but found it extremely difficult to work in this context, and looked to promote arts administration and marketing skills within the cultural sector as well as trying get the city administration to think more strategically. In this they worked closely with a growing number of educational and training organizations from North America and Western Europe, and closely linked to these, cultural policy professionals who aimed to intermediate between Western know-how and the realities of the Russian situation.

As well as supplying training programs these organizations and intermediaries also pushed for the commercialization of activities such as cafes, merchandising, loan of artifacts, renting of space, associated publishing. In the process they identified the following problems:

1 a lack of arts administration and marketing skills;
2 legal, bureaucratic, fiscal and cultural constraints on entrepreneurial activities;
3 lack of flexible human resources management powers (difficult to get rid of or financially reward staff);
4 a tendency for the large institutions to be self-contained and remote from other locally based cultural institutions;
5 the continued existence of many financially unviable small state institutions, over-manned, lacking basic skills and contributing little to the overall cultural life of the city.

Whilst the lack of professional managerial skills and the limited flexibility of human resources were important it was the local city administration which also proved an obstacle. The problems here were (Belova *et al.*, 2002):

1 the city's lack of a clearly outlined unified cultural and tourism strategy;
2 the opacity and clientele basis of its cultural funding system;
3 the wider lack of understanding of the economic dimensions and potential of the cultural sector;
4 the city's organizational confusion with regard to tourism, culture and small business development responsibilities.

Part of the task of the foreign foundations, organizations and consultants was to argue the case for the enhanced role of culture in the economic regeneration of the city – something Sobchak was clear about but which subsequent administrations have found difficult. The promotion of culture as a regeneration tool necessarily abuts on to wider policy domains which have rarely had to conceive of any systematic relationship to culture. For Russians culture was important, something precious that should have money spent on it – but as a gift to the patrimony *not* as economic investment.

The arguments for the economic role of culture, made by foreign-policy intermediaries but also by local reformers,[4] inevitably centered on developing the city's huge tourism potential and to the related search for new ways of attracting investment into the historic center. They looked to examples of Western cities which had somehow used culture to transform their image and regenerate the physical infrastructure of the city center. It was in this context that arguments about cultural industries appeared – as a contribution to debates over the cultural tourist "offer" and related investment in the historic center. Previous work (Landry, 1997; Causey, 2002) had established that St. Petersburg's tourist appeal was severely limited – its over-reliance on the huge prestigious institutions masked a lack of smaller, more innovative cultural activities which make up the ecosystem of a major destination city (St. Petersburg has its own *Rough Guide* and *Lonely Planet*!).

Cultural industries were thus introduced not in terms of large-scale global businesses but of supporting small, independent cultural producers who could somehow add to the overall cultural offer of St. Petersburg; and maybe even provide new funding possibilities for state supported culture. Further, the exclusive appeal to classical culture meant the image of the city was lacking in all contemporary profile. The absence of a small cultural producer sector also meant that the animation and retail/leisure occupation of the city centre either failed to appear or was completely inappropriate; newly refurbished areas lay empty, historic buildings were surrounded by shops selling things no "cultural tourist" would wish to buy. At root was the failure of small independent cultural producers to make any major impact on the landscape of the city – physical or imaginary. And behind this, it was argued, was the lack of any infrastructure or support for the development of cultural SMEs.

Culture and "transitional" economics

A fundamental debate ranges about the role of culture in post-Soviet Russia. In the 1990s, during years of economic collapse and chaos, the question arose as to what was the basis of state funding. What should it fund and how should it do so? Was it to remain in the framework of patrimony to be promoted and sustained by the state apparatus, as it had been up to 1989, or was it to find its way in the market place, somehow linked to the new consumption preferences of the population? At certain levels the state was not going to let go – it continued to have a central role in print and broadcast media regulation, and this has recently got tighter (Danks, 2001; Freeland, 2003). The issue of defending a national patrimony looms large, especially when linked to the nationality questions in the regions, where discussions are dominated by questions of the different rights of state, ethnicity and minority groups to preserve and promote their cultural heritage. This is not a debate restricted to policy elites – "high culture" has a widespread role in civil society and is seen as a crucial part of identity making, not only in giving a sense of national identity but also of personal survival in hard times (Causey, 2002).

This current project to promote cultural industries was set within a fairly clear narrative of Russia as a "transitional economy" where "expertise transfer" from the West would expedite the country's move from a command to a modern democratic and market society. In cultural-policy terms what happened in the West in the 1970s and 1980s – cuts in straight subsidy, professionalization, commercialization, diversification of markets and constituencies, deregulation and pluralism, etc. – was seen to be what would happen in Russia, albeit in a telescoped, maybe chaotic, manner. In terms of cultural industries this was linked also to the emergence of a small-business economy which was a key target of the World Bank and other development agencies (Leontief Centre, 1999).

However, it was quite clear that such a narrative was fiercely resisted in Russia, and especially so in the realm of culture. The schizophrenic history of Russian attitudes of "catching up with the West" or "finding a unique path" is too long to go into here, but these certainly continue to play a key role (Figes, 2002). The "shock therapy" administered to the Russian economy in the early 1990s produced a wave of disenchantment about the nature of transition. Many believed that the West would not let Russia catch up and simply assign it a peripheral role in the new world order. Thus market reforms were simply a means by which (mainly) U.S. capital would penetrate and dominate the assets and the markets of Russia. A more sweeping debate returns to the notion of Russia as not being part of "the West", but a different culture entirely. Politically this also looks to a strong nation state as a defense against globalization. The lack of openness and democratic freedoms which the West saw as crucial to modernization was, for this tendency, signs not of backwardness but of difference; and just as Western agencies got increasingly annoyed at the persistence of such attitudes many Russians embraced these as part of a national culture (Danks, 2001; Pilkington and Bliudina, 2002).

For those not concerned with cultural geo-politics there was still a burning awareness of the collapse of Russia's great power status, followed by loss of health care, education and science "as well as outward opulence against a background of growing poverty and moral degradation" (Pilkington and Bliudina, 2002: 8). The transition was frequently experienced as shame, disorientation, anger and despair. In this context culture was a key site of conflict – in which some national tradition or national soul was at stake. At the level of cultural policy the cut in jobs, pay and security, the fear of selling (out) a treasured heritage and the all-pervasive penetration of Western cultural goods also meant that, after early enthusiasm, the transfer of "expertise" could not be seen as merely technical but was hedged by ambiguity, cynicism and resentment.

St. Petersburg, of course, was always presented locally as "the most European city", the "window on the West'; so whilst not free from the narratives of Russia's Euro-Asiatic uniqueness, it was much more used to, and welcoming of, Westernizing narratives. However, as we saw, its self-image was closely associated with a classical European heritage; many saw the association of commerce and culture as something suspicious. This was never more clearly so than in the notion of "cultural industries". This opposition came not only from the administration and the large cultural institutions – who could afford to stress non-commercial idealism – but from those in the sector which the project had specifically set out to target – the independent cultural producers.

Independent cultural producers in St. Petersburg

This project first targeted the city administration, addressing workshops and seminars to key individuals and departments, as is the practice in the U.K. It became very apparent early on that the administration was not buying into this argumentation; it did not see culture as an issue other than to attract external funding – which it frequently siphoned off elsewhere; its various departments had little success in coordinating its own activities let alone cross-connecting to others on a "cultural" theme; and its hierarchical structures and frequent changes of personnel meant that it became impervious to external argumentation and rigidly frozen in fixed attitudes. Many international funding agencies had indeed tried to stress culture as a key factor of its economic future but became increasingly frustrated by its inflexibility, its incompetence and its corruption.[5]

It had also become apparent that any initiative led by the city administration would automatically be received with cold cynicism by those in the cultural sector. For the above reasons of incompetence and corruption certainly, but compounded by the experiences of the previous eighty years. Authorities were to be avoided, distrusted, lied to – sometimes ethically shunned. In such a context "partnership" was not an immediate option. In which situation the best solution seemed to be to work directly with the cultural producers themselves – in order to raise awareness of their potential role and indeed existence as a "sector," and to get them to take active ownership of any subsequent initiative. But in so approaching the sector – through research interviews, workshops, seminars, away days, along with many informal activities – it demanded a clarification of what exactly was intended by "cultural industries."

The terminology was a clear problem – as noted above a direct translation of the term evokes a factory, and not in any Warholian ironic terms either. Most of those interviewed initially rejected the term, until it was explained. But even when translated into "cultural business" it ran into a lot of problems, especially in a context where "businessman" could act as slang for "criminal" (often with reason; see Freeland, 2000). But the deeper territory of resistance was that of the boundaries between culture and commerce. This has a number of dimensions.

Initially it relates to the separation of subsidized culture ("art") and commercial culture ("entertainment") which was general in European cultural policy fields at least until the changes occurring there in the 1970s. This separation was firmly rooted in the cultural policy of the Soviet Union (Clark, 1995; Figes, 2002). However, this has a specific Russian dimension relating to the politicized role of "high culture" in the Soviet Union – when it was both a vehicle for political ideology but also a site for the transcendence of that ideology. The state tightly controlled cultural production for ideological purposes, but within a framework of respect for (usually pre-twentieth-century European bourgeois) "high culture". Soviet culture was of course defiantly non-commercial (i.e. non-capitalist). On the other hand, oppositional, "alternative" culture was by necessity non-commercial also; both in the sense that it did not generate income and in that it rejected immediate acceptance/popularity, envisaging for itself an ideal audience of the future (sometimes beyond the existing regime, or some more general "judgment of history," or a mixture of both). There was also a sense of an "international cultural mainstream" – which both officials and dissidents saw in terms of non-commercial "high culture." The former would produce artists of all kinds to compete in the mainstream to the glory of the nation; the latter would also look to this as some transcendent court of appeal, beyond the horizon of the nation. Unlike official culture, of course, the opposition espoused Modernism, but imagined in a cultural context very different from its actuality in the West – something made clear in 1989.

Oppositional culture had a sacred apartness; it was less about an Adorno-esque transcendence through the difficult and opaque work of art structured by a painful negativity, and more about the way Shostakovich and Akhmatova spoke a secret coded critique, a message of hope for those who could listen, keeping open the image of the artist in communication with the people which had long atrophied in the West. It was this that was threatened by Western commercial culture. The commercialization of culture was seen by oppositional culture as a degradation of culture; production for the market was as much (maybe more?) anathema for these as it was for official culture. The shocked encounter between a preserved (through political opposition) notion of "high art" and the reality of commercial culture industries in the West was prefigured in the reactions of exiled oppositionists (one thinks of Solzhenitsyn and Sakharov). It can be seen today in the cynicism of writers such as Victor Pelevin (Pelevin, 1999).

At a more basic level, as we have seen, the impact of the rapid market reforms in Russia was very hard on the cultural sector. Not only were the institutions and jobs of official subsidized culture under severe threat but the complex eco-system which had sustained oppositional culture (and the liminal areas which of course linked the two) in the form of university jobs, commissions from larger institutions, state grants, more private unofficial commissions, etc. – this eco-system suffered severely. With the result that there was frequently as much antagonism and anxiety about the collapse of state funding for culture on the oppositional side as there was the official side, at least for an older generation.

Cultural business, many said, equals cuts in subsidy to both institutions (and institutional jobs) and to individual artists; policies to develop the creative industries mean the sustainability of culture relying increasingly on the market. Culture produced for the market, it was said, means pleasing the lowest common denominator, creation of easy "entertainment" and the courting of immediate (and thus transitory) popularity. In this way, many people saw the autonomy of the artist – which should be at the heart of authentic cultural production – being betrayed "for thirty pieces of silver."

It was impossible for the project team to dismiss these fears as unfounded: pressures on state budgets, the power of the market, the globalization of the large cultural industries – these have set the context for cultural producers and policy makers in the West for two decades. As we saw, the simple idea of telescoped transition is frequently rejected as an impossible "catch-up" and seen as partly futile, partly a screen for Western penetration. In response we could only argue the following, taken from the project document:

> Like all changes, they present both dangers and opportunities.
>
> This is the situation of risk facing St. Petersburg culture today, whose addressing can be postponed, but not avoided. "Globalization" has involved accelerated flows of money, information, goods and people passing through cities. These flows include ideas, signs and symbols, the whole range of cultural products, information and ideas which make up the complex global cultural circuit of the contemporary world; and driven by publishing, satellite, internet, the internationalisation of production and distribution etc. which have transformed the immediate day to day context and wider significance of local cultures.
>
> Inevitably both local cultural production and cultural consumption now take place in a much wider context – culturally, economically and organisationally. All cities need to be much more reflexive and responsive to the changes – cultural policy therefore now looks to the preservation and promotion of local, place-based cultures through active engagement with these wider contexts.
>
> (Belova *et al.*, 2002)

The project thus initially presented itself to cultural producers in terms of the positive defense of a local culture in the face of a global economic and cultural threat. The direction offered was to engage with the economic dimension of culture through the promotion of independent cultural producers, now conceived not as artists but as freelancers and SMEs. In pursuing this line we hoped to establish a new set of understandings on behalf of both producers and policy makers for the development of support structures necessary for such a cultural ecosystem. SME policy was a key platform here – education and training, tax and business legislation, opening up of premises through managed workspace or incubation schemes, as well as managed "gentrification," small loan programs; and network development (Belova *et al.*, 2002).

The case for these policy tools, however, needed to be linked to an explicit cultural politics in order to avoid straight identification with a simple transition thesis. There was certainly an emphasis on the tourism, urban regeneration, employment and image benefits to the city – but these economic benefits were attached to a narrative of a "cultural renewal" of the city. The "cultural renewal" was about re-inventing St. Petersburg as the city of culture in a new century – a complex process but one that pointed to the wider development of a functioning civil society. Of course, this "modernization" could also be interpreted in terms of simple transition – where a market economy of small businesses and autonomous risk-taking symbolic professionals becomes the hallmark of a healthy emergent creative economy (Wang, 2003). A cultural politics in this context has to attempt the difficult task of making "modernization" work at the local (Petersburgian, Russian) level, and most especially if it is a narrative aimed at cultural producers whose sense of personal and professional identity is rooted in this context.

In short, the cultural industries argument needed to mobilize cultural and economic arguments in the context of a city culture. The linkage of this argument to the renewed infrastructure, image and identity of St. Petersburg reminds us that cities are divisions of labor but they are also an imaginative work (Blum, 2003). The cultural ecosystem is a key generator of economic value, but also of the imaginative work that is the city. Neither can thrive in a restricted civil society. The circulation of money, people, ideas, desires and things that is the city demands the spaces and places necessary to their flowing. This, at the present juncture, implies markets; not The Market, but multiple sites for a series of exchanges which are economic but also social and cultural.

Highlighting the absence of an SME support structure in St. Petersburg is one way of saying this; and the failures of this support structure are predicated on a wider failure of civil society. The catastrophe of "shock therapy" underlined the absence of any social and civil underpinnings of a market in Russia – and their absence precisely reveals the necessity of these underpinnings. Similarly, longstanding, culturally embedded structures of power continue play a role. Danks, for example, highlights the notion of the "hour glass society" (Danks, 2001: 193). In this notion, networking at the bottom and the top of the social structure are identified as extensive but horizontal – the elites effectively floating clear of civil society constraints and where legal and institutional support is barely trusted by those below, who rely on strong personal networks of trust. Both of these are destructive of the civil society context – the actions of the oligarchs speak for themselves, but networks based on personal trust quickly become restrictive, cliquey and, indeed, corrupt. The sort of fluid, open trust networks found in cultural ecosystems in the West are noticeably absent in St. Petersburg (O'Connor, 1999a, 2000a), where distrust of representatives, lead bodies, support agencies, intermediaries, etc. was a key finding of our research (CISR, 2003). People work with those they know and are wary of others.

The St. Petersburg project is in its second phase, which involves the establishment of a lead partnership agency and the placing of the cultural industry issue on the administration's agenda. But in trying to make these arguments, the "expert," as a suspect figure, has to find ways of gaining trust. Only to some extent can this be on technical terms, as some intrinsically valuable "know-how." If the cultural industries argument looks to an embodied performative shift or new habitus, then this not only takes time, but it is also done in a distinctly local context. Negus (2002), for example, characterizes intermediaries in the music industry in London as "public schoolboys"; in Manchester they have been seen as "Thatcher's children," the unemployed and disaffected. It involves the adoption of a "lifestyle," or more fundamentally a "habitus" in which the local cultural context is crucial. Thus despite the seemingly easy acceptance of Western cultural exports and models amongst the younger generation in Russia – and these are very much more open to the arguments around cultural industries – the selectivity of these has been strongly underlined by research (Pilkington and Bliudina, 2002). In this context it is quite likely that if the expertise is to be in any way effective it must engage on this local cultural terrain, which at once puts limits on the role of "expert" and opens up a new role of transnational cultural intermediary who has to have an explicit cultural politics.

Notes

1 The City of Helsinki (Urban Facts) was involved in the first phase of the partnership project and remain informally connected.

2 There were two phases to the project. First, an 18-month project (February 2001–July 2002) financed by the European Union's TACIS Cross Border Co-operation Programme, aimed at research and policy development for the cultural industry sector in St. Petersburg – conceived predominantly as SMEs and freelancers – as well as promoting entrepreneurialism amongst state- and city-funded cultural institutions. This was a formal three-way partnership between the city administrations of St. Petersburg, Helsinki and Manchester, though it was led and managed by Timo Cantel (Helsinki Urban Facts), Sue Causey (Prince of Wales Business Leaders Forum), Elena Belova (Leontief Centre) and myself at the Manchester Institute for Popular Culture. The second (January 2003–June 2004) was a continuation of this, within the TACIS Institution Building Partnership Programme (IBPP), under the specific line Support to Civil Society and Local Initiatives. This second program is a partnership between two independent not-for-profit agencies – the Creative Industries Development Service (CIDS), Manchester and a similar emergent organization in St. Petersburg. The intention is to establish this latter as a lead agency in promoting cultural industries in St. Petersburg, within a local policy context made more amenable to such initiatives.

3 See the film by Alexander Sokurov (2001) *Russian Ark*, Hermitage Bridge Studio – where the Hermitage becomes, by an historical paradox, the Ark where European culture preserved itself against the floods of ignorance and mass culture. Sokurov: "We are much closer to our past than Englishmen are to Victorian times. Our past hasn't become past yet – the main problem of this country is that we don't know when it will become past" (*Guardian*, 28 March, 2003, p. 3).

4 Such as Leonid Romankov, until last year the main cultural spokesperson for the City's Legislative Assembly; the Leontief Centre, the city's economic research agency; or Alexander Kobak, ex-head of the Soros Foundation and now head of the Likhachev Foundation, under whose aegis Charles Landry of Comedia, a leading international cultural consultancy, wrote the first cultural document pointing to the economic role of culture in the city (Landry, 1997).

References

Adorno, T. and Horkheimer, M. (1997/1944) "The Culture Industry as Mass Deception," in J. Curran, M. Gurevitch and J. Wollacott (eds) *Mass Communication and Society*, London: Edward Arnold.

Andrusz, G., Harloe, M., and Szelenyi, I. (1996) *Cities After Socialism: Urban and Regional Change and Conflict in Post-socialist Societies*, Oxford: Blackwell.

Belova, E., Cantell, T., Causey, S., Korf, E. and O'Connor, J. (2002) *Creative Industries in the Modern City: Encouraging Enterprise and Creativity in St. Petersburg*, St. Petersburg: TACIS-funded publication.

Berman, M. (1984) *All that is Solid Melts into Air: The Experience of Modernity*, London: Verso.

Blum, A. (2003) *The Imaginative Structure of the City*, Montreal: McGill-Queens University Press.

Bourdieu, P. (1986) *Distinction: A Social Critique of the Judgement of Taste*, London: Routledge.

Causey, S. (2002) "Cultural Institutions in Transition: issues and initiatives in Russia's cultural sector," unpublished paper presented at International Museum Association Conference, Salzburg, May.

Caves, R. (2000) *Creative Industries: Contracts Between Art and Commerce*, Cambridge, MA: Harvard University Press.

Centre for Independent Social Research (CISR) (2003) *Feasibility Study for Cultural Industries Agency St. Petersburg*, unpublished research.

Clark, K. (1995) *Petersburg, Crucible of a Cultural Revolution*, Cambridge, MA: Harvard University Press.

Danks, C. (2001) *Russian Society and Politics: An Introduction*, London: Longman.

Department of Culture, Media and Sport (DCMS) (1998) *Creative Industry Mapping Document*, London: DCMS.

du Gay, P. and Pryke, M. (2002) *Cultural Economy*, London: Sage.

Featherstone, M. (1990) *Consumer Culture and Postmodernism*, London: Sage.

Figes, O. (2002) *Natasha's Dance: A Cultural History of Russia*, London: Allen Lane/Penguin.

Florida, R. (2002) *The Rise of the Creative Class: and How it's Transforming Work, Leisure, Community and Everyday Life*, New York: Basic Books.

Freeland, C. (2000) *Sale of the Century: Russia's Wild Ride from Communism to Capitalism*, New York: Times Books.

—— (2003) "Falling Tsar," *Financial Times Weekend*, 1/2 November 2003: 1–2.

Hall, S. and Jacques, M. (1989) *New Times*, London: Lawrence and Wishart.

Harris, J. (2003) *The Last Party: Britpop, Blairism and the Demise of English Pop*, London: Fourth Estate.

Hesmondhalgh, D. (2002) *The Cultural Industries*, London: Sage.

Landry, C. (with Gnedovsky, M.) (1997) *Strategy for Survival: Can Culture be an Engine for St. Petersburg's Revitalisation?*, St. Petersburg: unpublished discussion paper.

Leadbetter, C. and Oakley, K. (1999) *The Independents. Britain's new cultural entrepreneurs*, London: Demos.

Leontief Centre (1999) *Rehabilitation of the Centre of St. Petersburg: Investment Strategy*, St. Petersburg: Leontief Centre.

Mellor, R. (1999) "The Russian City on the Edge of Collapse," *New Left Review*, 236: 53–76.

Negus, K. (2002) "The Work of Cultural Intermediaries and the Enduring Distance between Production and Consumption," *Cultural Studies*, 16: 501–15.

Nicolson, J. (1994) *The Other St. Petersburg*, St. Petersburg: n. p.

O'Connor, J. (1999a) *Cultural Production in Manchester: Mapping and Strategy*, Manchester Institute for Popular Culture, Manchester Metropolitan University; available online: <www.mmu.ac.uk/h-ss/mipc/iciss>

—— (1999b) *ICISS Transnational Research Report*, Manchester Institute for Popular Culture, Manchester Metropolitan University; available online: <www.mmu.ac.uk/h-ss/mipc/iciss>

—— (2000) "Cultural Industries," *European Journal of Arts Education*, 2 (3): 15–27.

O'Connor, J., Banks, M., Lovatt, A. and Raffo, C. (1997) "Modernist Education in a Postmodern World: critical evidence of business education and business practice in the cultural industries," *British Journal of Education and Work*, 9: 19–34.

—— (2000a) "Risk and Trust in the Cultural Industries," *Geoforum*, 31: 453–64.

—— (2000b) "Attitudes to Formal Business Training and Learning amongst Entrepreneurs in the Cultural Industries: situated business learning through "doing with others," *British Journal of Education and Work*, 13: 215–30.

O'Connor, J. and Wynne, D. (1998) "Consumption and the Postmodern City," *Urban Studies*, 35: 841–864.

Pelevin, V. (1999) *Babylon*, trans. A. Bromfield, London: Faber and Faber.

Pilkington, H. and Bliudina, U. (2002) "Cultural Globalisation: a peripheral perspective" in H. Pilkington, E. Omel'chenko, M. Flynn, U. Bliudina and E. Starkova (2002) *Looking West?: Cultural Globalization and Russian Youth Culture*, Pennsylvania: Pennsylvania University Press.

Volkov, S. (1995) *St. Petersburg: A Cultural History*, New York: The Free Press.

Wang, Jing (2003) "The Global Reach of a New Discourse: how far can "creative industries" travel?," *International Journal of Cultural Studies*, 7: 1.

4 Putting e-commerce in its place

Reflections on the impact of the internet on the cultural industries

Shaun French, Louise Crewe, Andrew Leyshon, Peter Webb and Nigel Thrift

Introduction

In this chapter we explore what happens in theory and in practice when Internet technologies are adopted within the cultural industries. This dual focus on e-commerce and the cultural industries is, we argue, significant not just empirically but conceptually too. Whilst there has been considerable debate about what precisely constitutes the cultural industries (see Du Gay and Pryke, 2002, for example), there is a reasonable consensus, as Power and Scott in the introduction to this volume make clear, that at the very least their key attributes include an involvement in commodities and services with high levels of symbolic content. We argue that while the Internet has the potential to reconfigure a whole host of industries, study of the impact upon the value chains of those with high levels of symbolic content is particularly important given the Internet's potential to reconfigure the very way in which we think about and define the cultural industries.

A key concern in much recent literature has been to explore the capacity of e-commerce to disintermediate – or, more accurately, *reintermediate* (French and Leyshon, 2004) – production networks and value chains within established industries. That is, to short-circuit prevailing lines of connection within networks of production and consumption. The Internet is seen to offer considerable potential for the reduction of both entry and fixed costs, and can thus disintermediate existing retailers whilst also offering access to a much larger customer base. In addition, the Internet offers the potential to circumvent, re-order or even remove what were hitherto key nodes within prevailing value chains, and to directly connect consumers with producers and designers/creators, or even with each other. Customers in turn have access to far greater information about commodities, their value and their supply chains. Not only does the Internet have the potential to connect every node within a value chain, but it also makes it possible for entirely new intermediaries to enter value chains and, in extreme cases, to remove certain nodes altogether.

In this chapter we consider the impact of the Internet upon three specific sectors. Two of them are prototypical cultural industries: fashion and music. The third sector is retail financial services, which is clearly not a cultural industry, at least not in terms of traditional definitional conventions. However, we unapologetically include an analysis of the financial services industry within this chapter because it serves as an important comparator and context for the analysis of the two cultural industries. Although we argue that the commodity form and the topologies of individual value chains matter, the analysis also reveals that in many respects the impact made by the Internet and e-commerce upon the cultural industries is not that distinctive, and should be seen as part of a broader set of changes that are generic and sweeping across all industries, regardless of whether they are defined as cultural industries or not.

The remainder of the chapter is organized into three main parts. In the next part we discuss in more detail the theory of Internet disintermediation and the concept of the value chain. We then go on to trace out the topologies of the value chains of the music, fashion and retail financial services industries. The third part of the chapter considers the organizational and spatial impacts of e-commerce upon the sectors in question. Despite the difficulties of undertaking an evaluation of the impact of such a relatively new technology, a number of key issues emerge from our comparison of the music, fashion and retail financial services industries. The final and concluding part is given over to a discussion of these issues.

Placing networks: the spatial logics of the music, fashion and retail financial services value chains

At the crux of arguments surrounding disintermediation is the belief that the architectures and geographies of value chains across a range of industries are being undermined and reconfigured by the emergence of the Internet and the possibilities of e-commerce (French and Leyshon, 2004). The concept of the value chain has emerged in response to attempts to understand the organization of global corporations, and their ability to co-ordinate their activities over great expanses of space-time (Dicken, 2003; Gereffi and Korzeniewicz, 1994). In many industries, a significant proportion of the value chain is dedicated to material logistics, and the movement of physical commodities between sites of production and consumption. However, the advent of e-commerce has led many commentators to question the role of extant material value chains. In particular, commentators like Evans and Wurster (1997, 1999) have argued that the informational qualities of certain commodities (such as music and financial products, for example) make them reducible to the zeros and ones of electronic communication. In addition, e-commerce is also seen to offer consumers the potential to intervene within the value chains of a wide range of sectors in new ways. However, these claims have not gone unchallenged (see Liebowitz, 2002), and a good deal of hyperbole – or, as Woolgar (2002) puts it, cyberbole – surrounds the projections of what difference the Internet makes to economic exchange (Leyshon *et al.*, 2004).

This chapter reports on research that has sought to determine what happens to the organizational structures of industries after the introduction of e-commerce. It draws upon findings from a two-year ESRC-funded project that involved interviews with key informants in firms within the music, fashion and retail financial services industries in Europe and North America. But before we assess the current and future impact of the Internet we need first to examine the contemporary spatial fixes of these three industries.

For heuristic purposes, we conceptualize all three industries as being constituted of value chains that are organized as networks. From this perspective commodities can be thought of as flowing through industry value chains from networks of creativity, to networks of reproduction, distribution and consumption (see Leyshon, 2001). As we shall see, the conceptualization of value chains in terms of networks of creativity, reproduction, distribution and consumption provides a powerful analytical tool for examining the impact of the Internet, but it is not without its limitations. These networks are relationally constructed and reproduced dialectically via nodes and agents. Thus, the flows of information, commodities and people through such value chains are rarely the types of linear shifts in time-space suggested by the language of creation, production, distribution and consumption. Despite these limitations, the tracing of such network topologies not only reveals the distinctive geographies, histories and temporalities of the music, fashion and retail financial services sectors, but also the differential vulnerability of each to the possibilities of disintermediation through e-commerce (French and Leyshon, 2004).

The fashion value chain

Fashion retailing is constituted by a distinctive set of interconnecting networks of production, promotion, sales, consumption and regulation. These networks are spatially fragmented and exhibit markedly asymmetrical power relations. The fashion industry is positioned within a nexus that connects design, promotion and display with clothing production and retailing. This nexus is highly differentiated; it works itself out in different ways to produce and sustain multiple readings of the fashion industry, which depend on very different geographies of design, clothing production, promotion and retailing (Crewe, 2003; Entwistle, 2000). The sector is markedly segmented and is striated along organizational and quality lines.

Networks of creativity and design exhibit distinctive spatial structures, centered on four key cities: Paris, Milan, London and New York. The organizational ecology of networks of creativity and design is complex and conventionally conceived as bi-polar, characterized on the one hand by small, fledgling, independent designers, often recent graduates from fashion institutions who attempt to "go it alone" (see McRobbie, 1998 and Purvis, 1996 for examples), and on the other hand by large organizations such as LVMH and Prada who wield considerable amounts of market power and control.

Networks of production in the fashion industry are also segmented and striated. There are two dominant spatial fixes at work here. First, spatial agglomeration as illustrated in the case of garment production clusters in New York City,

London's East End and the Sentier district in Paris, for example. All three are important centers of production characterized by small, highly flexible and responsive production units employing poorly paid, often immigrant labor (Green, 2002; Rath, 2002; Zhou, 2002). Volatile fashion demands and a marked division of labor encouraged manufacturers to seek proximity to fabric suppliers, cutters and ancillary suppliers. The second production spatial fix is more global in orientation. Global supply chains have been a long-established feature of the fashion industry (Elson and Pearson, 1981; Phizaklea, 1990; Wright, 1997), with Export Processing Zones (particularly in the Far East and, most recently, China) being particularly important sites of garment manufacture. Despite the labor-cost benefits of such overseas production sites, the rapidly changing fashion cycle and, in particular, the requirements for rapid response and "fast fashion" have made womenswear in particular less prone to spatial relocation (Entwistle, 2000). The third fashion network is that associated with promotion, distribution and dissemination. It comprises cultural intermediaries involved in the selection and promotion of fashion (such as fashion editors, photographers, journalists and the like), who exercise control over the dissemination of fashions through the global media. Typically the production of fashion has been thought about in terms of the "trickle down" of styles, from catwalk to high street (from designer to mass-market retailer). In part it is clear that this dynamic is still at work, as is evident through the twice-yearly staging of world fashion weeks showcasing the coming season's styles and concepts. It is at such events that the different geographies of fashion unfold and are connected through the nexus of London, Paris, Milan and New York – spaces for the display, performance and enactment of "fashion." These are highly charged and influential meeting grounds for a number of actors who shape fashion consumption, including designers, the collections themselves, supermodels, media pundits, photographers, magazines and the trade press.

The spatial fixes of design, production and promotion are, in turn, intimately connected to the economics of the fourth network, that of clothing retail. We shall illustrate this by specific reference to fashion consumption within the U.K. The retail nodes within fashion networks are striated and divided along quality lines, and include the *haute couture*, *prêt-à-porter* (ready to wear) and diffusion ranges of key design brands such as Armani, Chanel and Gabbana at the upper reaches of the market; design-led chains such as Kookai, Whistles and Jigsaw; middle-market chains such as Marks & Spencer and Next; and low-end stores such as Top Shop and Zara whose design impetus and "fast fashion" model is proving commercially very successful (D'Andrea and Arnold, 2003). From a situation in the 1980s where six middle-market clothing retailers dominated the U.K. high street, the early twenty-first century has been characterized by declining levels of concentration and corporate control. Growth is strong at both the upper and the lower reaches of the market: on the one hand, the low-cost, high-fashion discount end (ranging from firms such as Matalan, Bay Trading and New Look to Top Shop and Zara) and the design-led, quality end (examples here include Jigsaw, Diesel, Ted Baker and French Connection) at the other. These two nodes of growth in turn reveal distinct spatialities, in terms of both production and consumption. The discount retailers favor low-cost foreign production locations and cheap high-street or out-of-town retail locations, whilst the design-led retailers favor domestic design input, higher-cost European production locations and specialized retail spaces, often in cultural quarters of cities, with high rental costs.

The music industry

As in the case of fashion, value production within the music industry has conventionally been constituted by a set of overlapping and interconnecting networks; networks of creativity; networks of reproduction; networks of distribution; and networks of consumption (Leyshon, 2001, 2003; Leyshon *et al.*, 2004). Although these different networks are constituted by a diverse range of actors, institutions and spaces (artists, producers, talent agencies, recording studios, performance venues, retailers, and so on), what is striking about the music industry is the way in which record companies, and in particular the "big five" conglomerates (AOL-Time Warner, Sony, Bertelsmann (BMG), Universal-Vivendi and EMI), have been able to dominate the music industry value chain over a relatively long period of time. Record companies have been able to position themselves as obligatory passage points (Callon, 1991) through which cultural material needs to flow if it is to be commodified in such a way that maximizes the creation and extraction of economic value. That is, music is stabilized and transformed into compact discs (CDs) and other media that, through the mobilization of the considerable marketing powers of such companies, can generate significant consumer demand and sales in the world's major consumer markets. From the making and performance of music within networks of creativity, to its stabilization within networks of reproduction, through to its dissemination to consumers within networks of distribution (the promotion and marketing of such physical media, and its distribution to retail outlets), record companies play a key role. Record companies control the most important cultural and economic assets within networks of reproduction and distribution and they seek out, foster, manage and fund talent within networks of creativity.

The dominance of record companies within the music industry value chain is reflected in the spatial logic of the industry. The offices of the big five recording companies are concentrated in New York, Los Angeles and London, the world's leading centers for popular cultural production. Such centers also host offices of many other institutions integral to the reproduction of music (such as publishers, lawyers, artistic management, the music press, and so on). In this sense these cities operate, in the language of actor-network theory, as music "ordering" or interpretative centers, gathering in, sorting, selecting, valorizing, commodifying and reproducing cultural material. So, although networks of reproduction, distribution and consumption have very different spatialities (Leyshon, 2001), it is the spatial logic of networks of creativity that embeds record companies and associated reproductive institutional milieux within global cultural capitals, which are at the heart of the music industry value chain both in terms of the production and extraction of economic value. This said, two other spatialities are worthy of particular note. First, the shifting landscape of creativity. In comparison with networks of reproduction, networks of creativity are more institutionally fluid. Nevertheless, it is possible to identify at least two important characteristics. First, the popular, global music industry has been characterized by the domination of the creative output of a very small number of countries, in particular the United States, the U.K. and Jamaica (Connell and Gibson, 2002; Power and Hallencreutz, 2002). Second, with the possible exception of the musical cultural capitals of New York, Los Angeles, London and Kingston (Jamaica), the geography of musical creativity in these countries is quite dynamic and fluid, different centers of creativity being significant at different times and in different places. In the case of the U.K. market, for instance, the prominence of the Manchester sound in the early 1990s was initially superseded by American Grunge, centered on Seattle, then in the mid-1990s musical interest shifted toward Bristol and a loose amalgam of "trip hop" artists, and most recently the contemporary dominance of dance music has seen attention switch once again to artists associated with places such as Ibiza and Ayia Napia.

The second spatiality of particular significance is that associated with distribution. As with networks of creativity, networks of musical distribution are more diffuse and decentered than those of reproduction. Nevertheless, just as record companies have risen to dominate networks of reproduction, distribution and creativity, large high street (and increasingly shopping mall) retailers have emerged which account for a large proportion of music sales (such as Virgin and HMV, for example). However, as we shall discuss later, the dominant role of such large retailers, and the spatial logic underpinning this dominance, within networks of distribution is threatened directly by the Internet and e-commerce.

The retail financial services industry

Turning, finally, to the case of retail financial services there are similarities between the manner in which record companies have been able to dominate the music value chain and the manner in which financial services firms have sought to strengthen their position *vis-à-vis* consumers, through the mobilization of information and communication technologies (ICTs). Although engaged in seemingly very different sectors, both record companies and financial services firms have sought to dominate and control the selection, sorting and interpretation of key information, and thus their respective value chains. Within the U.K., for example, the introduction of ATMs, telephone banking, credit cards, debit cards and credit-scoring technologies has enabled banks and building societies to reconfigure the traditional knowledge structure of the value chain to their own advantage. In particular, such technologies have provided firms with new tools with which to tackle the perennial problem of information asymmetries, which exist between providers and applicants for loans, insurance policies, mortgages and the like (Leyshon and Thrift, 1999). First, these technologies enabled firms to assemble and centralize customer behavioral data in a much more systematic and codified way than ever before. Second, they provided lenders with new tools with which to analyze and compare their own customer data with data held by other lenders, retailers and state institutions. Third, by so doing they enabled banks to hollow out and close branches, switching much of the decision-making and transactions activity into centralized units. In turn, this has enabled banks and building societies to reduce their branch structure and concentrate key decision-making activities in regional financial centers (French and Leyshon, 2003). At the same time global financial centers such as the City of London continue to play crucial roles in the industry value chain, being the location of the registered offices of many retail financial services firms, as well as hosting regulatory institutions, industry and professional bodies, financial markets and the financial press. As such, financial centers such as the City of London act as a kind of financial services *cultural* centre (Thrift, 1994).

The hollowing out and contraction of the branch networks of retail financial services firms within the U.K. has been facilitated by, and in turn has helped to further encourage, the increasing commodification of retail financial services products. Whilst the combination of credit-scoring and more at-a-distance distribution systems have helped banks, building societies and insurance companies to cut costs by centralizing decision-making and strengthening their position *vis-à-vis* consumers, these forces have also facilitated the decoupling of consumption from branch networks, thereby transforming knowledge that was once proprietary into a more or less common good. This has significantly reduced the barriers to market entry within the retail financial services industry and eroded the inertia within the market that traditionally tied customers to individual firms. As a consequence, financial services consumption has become much more decentered, customer service has become more of a key issue, and new providers have entered the retail financial services market. New non-traditional providers, such as Virgin,[1] and retail chains, like Marks & Spencer and Tesco,[2] have been able to successfully enter the market through the leveraging of well-established brands, as well as established distribution networks, in a market place where brand, for the reasons previously outlined, has become an increasingly important asset.

Having outlined the prevailing organizational and spatial logics of the three industries, in the next section of the chapter we move on to consider the organizational and spatial impacts of e-commerce upon them.

Value chains after the internet

In some respects attempting to undertake a full evaluation of the impact of e-commerce on the fashion, music and retail financial services industries is premature. E-commerce is still a relatively new technology, so a full account of its consequences cannot be properly accomplished for many years. However, it is clear that e-commerce has already had important impacts upon industries, and that the degree to which it has already stabilized as an accepted way of doing business varies markedly. This differential impact was borne out in the three case study industries, as we illustrate below.

Fashion

One of the reasons that the Internet and e-commerce generated such excitement among a range of economic commentators was its inherent potential to bring about a shift in power from market incumbents to relatively new market entrants. This was to be achieved through the disruption and reconfiguration of established industrial networks, through the development of new forms of electronic connection that enabled new ways of doing business and extending market reach. Within the fashion industry, the appeal of e-commerce was the potential it offered to undermine the power of large retail outlets, distributors, intermediaries and the designer hubs within the value chain. In so doing, it held up the promise of a more "empowered" consumer, one who would be able to sift and sort through an enormous array of fashion information in a very short space of time, one who would have instantaneous access to emergent fashion trends from around the world. This offered the potential at least for new forms of democratized consumption and for the subversion of corporate monopoly over supply and distribution chains. Early forays into Internet fashion retailing were not auspicious however. Companies such as Boo.Com were amongst the earliest, and most spectacular, fall-outs from the dot.com downturn, in no large part due to inflated financial valuations and expectations (Cassidy, 2002; Malmsten *et al.*, 2002). But, more specifically, such prototypical fashion Internet sites were slow, revealed insufficient garment and fabric detail and led to consumer concerns over quality, cut, size and fit.

More recently, there are signs that Internet fashion retailing might be impacting more significantly on particular segments of the market. First, there has been considerable success at the bottom end of the market where a number of Internet sites have successfully reduced overheads and offered mass-market products at discount prices. Second, a number of companies with strong and established brand presence on the high street have successfully made the move into Internet retailing on the basis of the brand (such as Levi's, for example). This strategy sells the product through reputation and trust. A third group of firms have moved into e-commerce on the basis of expertise in other forms of long-distance selling, such as mail order (as have Boden and Racing Green, among others).

Fourth, a number of companies have tapped niches that high-street firms have routinely ignored. Thus, some social groups have often been marginalized by conventional retail formats and many high-street retailers. For example, the inherently embodied nature of clothes shopping makes it a troubling exercise for certain groups of bodies (such as the large, the pregnant and the old (Bruzzi and Church-Gibson, 2000)). Shopping at a distance in the privacy and comfort of one's home removes the need to publicly confront and expose one's body. In this we can see how cyber-identities are potentially freed from their anchorage in off-line bodies and are capable of being imaginatively or discursively reconstructed. None of which, of course, is particularly revolutionary. But there are recent signs that the Internet is having an altogether more dramatic and interesting effect upon the business operations of a number of new economic actors.

Emergent second-generation Internet fashion sites are aiming to go beyond and improve upon conventional high-street shopping rather than replicate it. They are strategically and creatively re-ordering and re-working both the spatiality and the temporality of fashion's value chains and, in so doing, are beginning to re-write the rules of fashion supply in a number of interesting ways. Not only are such sites far more technologically adept, with more pictures and color, but they include catwalk and collection coverage and personalized 3D mannequin services (see, for example, Archetype in the U.S.). The emergence of the "keyboard brand" is another concept that is enabling the flattening of fashion's traditional value chain. The designer, brand or retailer here becomes the keyboard or the screen from which customers can download what they want. In the U.K., Oki-ni (www.oki-ni.com) is perhaps the best example of this, where customers can see single pieces of customized clothing on display or via terminals in-store or on hand-held devices. They can learn more about the labels, search through the store's archives and place an order, safe in the knowledge that their product will be one of only perhaps fifty such items in the world. By removing the point-of-sale Oki-ni actively displaces the assistant–customer relationship and reveals the ambiguities of fashion shopping, consumption and sale in the material and virtual worlds. The Oki-ni model is significant in consciously removing a number of nodes that exist under conventional models of fashion production and supply. In particular, Oki-ni distributes a range of products and brands from one central storage node, directly to the consumer. And whilst this makes sense in terms of the economics of warehousing and logistics, its real radical potential lies in the way it uses the Internet to transform the geographies of commodity supply and fulfillment, stripping out former key nodes such as local storage and distribution centers, serving regional retail outlets with large numbers of sales staff. What the Oki-ni model emphasizes above all, is the complex ways in which the virtual and the physical worlds of fashion are entwined and connected.

Nevertheless, at present the impact of the Internet on the fashion sector is much less well developed than in the financial services or music industries. The fashion industry is still heavily reliant on the unique nature of the commodity form: touch, fit, fabric, quality and material aesthetics all matter in ways that mean consumers on the whole tend to prefer to be co-present with the clothes they buy. The fashion industry is also locked into both a temporal cycle of seasonality and into key display and performance spaces: the flagship store, the international collections and Fashion Weeks. The fashion design industry tells a story that is still firmly located in particular global cities such as Milan, Paris, London, New York and increasingly Tokyo, places that have particular fashion histories and cultural associations. And within these global cities there are also micro-geographies at work: in London the distinctions between the performative selling spaces of Bond Street and the more hidden production spaces of Brick Lane are clear examples here.

Therefore, fashion consumption as event and experience remains important, and it is far more difficult to reduce this to a momentary technological transaction. The problem, we argue, with existing literature on e-commerce is that it tends to reduce consumption to the moment and means of purchase, and in the case of fashion presumes this to be a de-sensory experience. This approach fails to acknowledge fashion consumption as event, display and performance, where the sounds, the stage, the props, the feels and the fit are all important. It is rarely, if ever, a disembedded, disembodied rational exchange of money for commodity, but rather demands co-presence. The physical and material qualities of the commodity matters profoundly here. And we might thus speculate that e-commerce is therefore not disrupting and reshaping the existing geographies of fashion, which remain stable, accentuated even. It is, rather, building on and reducing existing geographical, organizational and market divisions.

The music industry

In contrast to the fashion industry, the Internet appears, certainly at first glance, to be having a much more significant impact upon the traditional geographies of the recorded music industry value chain (Leyshon, 2001, 2003; Leyshon *et al.*, 2004). Whilst music is itself inherently variable and complex, the outputs are overwhelmingly in the form of mass-produced standardized forms (CDs and records, for example). Because of the largely standardized and commodified nature of its outputs the music industry is susceptible to reintermediation through the Internet in at least two ways. First, through the mobilization of the Internet as a distribution network for traditional music formats. In this case the Internet acts to undermine the traditional spatial logic of the industry, which has enabled the large record retailers such as HMV and Virgin Megastore to dominate distribution networks. In much the same way as, we argue below, the Internet works to accentuate the reconfiguration of the role of the branch network in the financial services sector, the establishment of successful Internet-based distribution operations by Amazon and companies like HMV themselves heralds, at the very least, a similar reconfiguration of the role of high-street record shops.

More fundamentally, the Internet and attendant technologies offer the potential for the *distribution and mobilization of very different forms of music formats.* In particular, they permit the production, distribution and consumption of music in new digitally compressed formats such as MP3. The Internet makes it possible to send music electronically rather than in the hard material forms that have dominated to date. These processes of reintermediation have made the recorded music industry into a *much more contestable space*, increasing competition and threatening the record companies place at the heart of the music value chain.

Indeed, the impact of the Internet within the music industry has acted as a "tipping point" that has brought about a broad recognition that the industry needs to undergo substantial reorganization (Leyshon *et al.*, 2004). As a result, the early years of the twenty-first century has witnessed a range of experiments with new Internet-oriented business models (Lucas, 2002) within the music industry to try and adapt to the problems of distributing music in digital form while, at the same time, keeping control over the copyright of music upon which the reproduction of the music industry has been built (Leyshon, 2003; Power and Jansson, 2004). They range from relatively modest attempts by the large record companies to build in a digital distribution capacity while simultaneously offering strong support to legal measures to combat the threat posed by Peer-to-Peer networks, through experiments with subscription-based downloading services, to attempts to completely reintermediate the music industry value chain. It is noticeable that it is on the edges of the music industry that the more radical attempts to accommodate the demand for music in digital form have emerged. Thus, while firms such as Napster, Kazza and Morpheus illustrated the potential market for downloaded music, it was the computing firm Apple, rather than a record company, that pioneered the first successful and legal large music download business model. As we have argued elsewhere (Leyshon *et al.*, 2004), the music industry has in large measure responded to the challenge of the Internet by adopting strategies that seek to defend and support business models that seem better suited to the twentieth than the twenty-first century. One of the reasons for this, we argue, is due to the highly precarious nature of executive employment within the music industry, which discourages the industry to embark upon long-term processes of restructuring as those that initiate such measures are rarely around long enough to reap their rewards. Thus, much as Feng *et al.* (2001) have argued for the "new economy" in general, the music industry has used the capital markets as a form of business strategy incubator, monitoring start-up firms developing new music industry business models, attacking those with legal sanction that are deemed to transgress laws in intellectual property rights, and buying up those that seem to offer legal and durable solutions to the distribution of music in digital form.

The financial services industry

As previously discussed, the commodification of financial products and the introduction of ATMs, credit scoring and other information and communication technologies have already had a significant impact upon traditional networks of production, distribution and consumption within the financial services industry. One possibility is, therefore, to think of the Internet as simply the most recent manifestation of a long-term process of ICT-driven restructuring within the industry. As such, the impact of the establishment of e-banking operations, Internet-based insurance sites, Internet brokers and the like can be thought of as parallel developments to the introduction of technologies such as ATMs. That is, within the financial services industry the Internet is but one of a number of forces driving the "hollowing out" of branch networks and thereby opening up retail financial markets to new competitors. In the wake of the collapse of the dot.com bubble, and the associated re-evaluation of the intellectual inflation surrounding e-commerce, such a position has been characteristic of much of contemporary discourse in the industry, both in the U.K. and the U.S. As it has become clear that the majority of customers have not, as widely predicted, utilized the Internet *instead of* but in *addition to* the branch and the telephone, e-finance has become less of a strategic subject in and of itself within the industry. Contemporary discussion of e-banking and the like has instead tended to be increasingly framed within new debates concerning "multi-channel distribution" or "Customer Relationship Management" (CRM), for example.

In contrast, the Internet and the emergence of e-finance has been framed in much more revolutionary terms by a second body of literature. This second, alternative perspective stresses the technology's potential to fundamentally reconfigure existing financial services' value chains. Evans and Wurster (1999) argue, for example, that the Internet is undermining the traditional hierarchy of banking and in its place a new banking "hyperarchy" is emerging. More particularly, Evans and Wurster argue that the traditional linear model of interaction and value creation between banks and customers – via transaction processing centers, ATMs, branch counter staff and other conventional nodes – is being rapidly superseded by a reconfigured business model, allowing for much more complex and differentiated production, distribution and consumption configurations. The erosion of the role of the branch as a significant barrier to market entry through the introduction of ATMs and telephone banking services has, as we emphasized earlier, already enabled supermarket chains and others to enter the retail financial services market. However, Evans and Wurster (1997, 1999) argue that the Internet offers the potential to radically undermine the hold of traditional financial institutions over the retail financial services market. The emergence of Internet search engines, personal financial management software packages, such as Quicken and Microsoft Money, and specialist Internet-based financial information-brokers, such as Motley Fool (www.fool.co.uk), are seen to have significantly amplified the deconstruction of traditional industry knowledge hierarchies. In so doing existing tensions around the production, mobilization and interpretation of knowledge have also been significantly heightened. On the one hand, the Internet offers financial services companies potential access to much richer customer data, data that can be used, in conjunction with credit-scoring technology and intelligent networked software, to further strengthen the position of traditional providers *vis-à-vis* customers. Thus significant potential exists for financial services companies to develop much more sophisticated technologies of consumer discrimination, with new temporal and spatial dimensions (such as customer profiles that are constantly updated and temporally sensitive in the case of the former, and new landscapes of financial exclusion, financial inclusion, financial sub-inclusion and financial super-inclusion in the case of the latter).

On the other hand, just as the record companies' dominance of music industry value chains is threatened by new modalities of music creation, reproduction, distribution and consumption, so the Internet has the potential to significantly disrupt value chains within the financial services industry, undermining the central role of banks, building societies and insurance companies. There are at least five ways in which the Internet poses a threat to traditional providers. First, the Internet allows, even more than the telephone, new providers to make rapid and relatively cheap entries into financial services markets. Second, many of the new opportunities for the production, interpretation and mobilization of financial information, which the Internet facilitates, are to do with the new actors and intermediaries previously highlighted. Third, the deepening of a financial services "information ecology" (Nardi and O'Day, 1999) provides consumers, or more properly particular groups of consumers (that is, those who are relatively wealthy, financially literate, Internet savvy and on-line) with significant new opportunities to overcome traditional consumer inertia and more actively differentiate and discriminate against financial services providers. Fourth, in addition, the Internet also provides new opportunities for credit-scoring companies and financial software providers to leverage their place within the landscape of financial knowledge. Fifth, although there is still a significant gap between rhetoric and practice, developments in the field of e-money, micro-payments, e-purses and the like also offer new Internet actors, such as Internet retailers and Internet Service Providers, opportunities to provide services which up until now have been the traditional provenance of the bank, insurance company and building society. Thus, the Internet has the potential to offer financial services companies opportunities to further strengthen their place within information hierarchies and to reduce costly branch networks, but it also has the potential to threaten the role of traditional providers as the hub of the industry through the empowerment of key consumer demographics, the introduction of new "info-agents," the further erosion of barriers to market entry, opportunities for the introduction of new types of money and the enhancement of the role of existing financial information intermediaries.

However, close examination of the retail financial services sector reveals a rather more complex picture. Despite earlier predictions of their imminent demise, traditional providers seem to have fared remarkably well. Whereas in the case of the music industry new types of music format have very rapidly been mobilized, adopted and institutionalized, the development of new forms of e-money and e-purses remains at a preliminary stage. In the U.K., for instance, it has only been very recently that the necessary regulatory framework for the issuance of e-money has begun to take shape (see FSA, 2001, 2002). The comparatively slow development of e-money is due to the role of trust in constituting money. Not only do consumers have to trust this new form of money and its issuers, but financial regulators, such as the Financial Services Authority in the U.K., and banks, credit-card companies and other existing credit and debt providers must also ensure the trustworthiness of any new money, for a crisis of trust in any part of the monetary system could quickly spread to endanger the stability of the financial system as a whole. Thus, whereas in the case of the music industry new regulatory systems have begun to be developed in *response to* their disruption by the widespread take up of new digital music formats by consumers, the existing regime of monetary issuance has been extended by the FSA, in consultation with existing and potential issuers of money, to create robust and trustworthy structures to enable a subsequent mobilization and take up of e-money.

Not only have new forms of money so far failed to have the impact on the financial services sector that software formats have had on the music industry, but the Internet's impact on traditional financial providers and networks of production, distribution and consumption has also been far less revolutionary than many earlier commentators predicted. Whilst cursory evidence suggests that branch network restructuring continues apace, and that the Internet has certainly provided an important additional factor in the rationale for such restructuring, we do not appear to be witnessing the death of the high-street branch, at least not yet. While care must be taken not to read too much into bank marketing campaigns, NatWest bank's recent high-profile advertising campaign in the U.K. – which stressed its commitment to branch banking – reflects the continued attachment of consumers to branches as service points and as symbols of organizational reliability. Nevertheless, it would be now equally hard to imagine a major bank successfully operating in the U.K. or the United States without providing Internet banking services. In addition, a significant number of "Internet banks" – such as Smile, Egg, Intelligent Finance, for example – have managed very successfully to become established in the U.K. market place over the last four or five years. There is little doubt that part of the success of such banks can be attributed to the development of independent Internet brands. However, they have also undoubtedly benefited from their status as spin-offs from existing financial services firms. Smile was established as a spin-off from the Co-op bank, Egg from the Prudential Insurance Company and Intelligent Finance from Abbey National. Thus, these Internet banks have also greatly benefited from the expertise and assets – such as money, labor, infrastructure and trust – that they have been able to draw upon from the parent organization. In contrast, high-profile attempts to create stand-alone Internet banks from scratch – such as e-finance in the U.K. and Wingspan in the U.S. – have been unsuccessful.

To date then, we would argue that the story of the Internet and the retail financial services sector has been one that falls somewhere in between that of the fashion and music industries. The importance of trust, coupled with the consumer's continued attachment to high-street branches and the relative success traditional providers have had in translating (Callon, 1991) the Internet, has meant that the Internet has not yet threatened the place of the bank, building society or insurance company in the same way as it has the record company. However, at the same time financial services do not have the same materiality as clothes, so whilst it is as yet uncertain how far virtual mannequins, pop-up "fabric boxes" and the like will be able to displace the importance of "being there" in acts of fashion consumption, it is safe to conclude that the financial services value chain remains vulnerable to further reconfiguration. Internet tools such as mortgage calculators and info-mediaries like Motley Fool have undoubtedly made retail financial services more transparent and competitive, empowering certain groups of consumers in the process, whilst Internet banks have shown that it is possible to inculcate trust online and e-money and e-purses seem finally to be on the horizon, for example. As such, it is possible to read the "soft" account of the Internet currently in circulation within industry circles as a very particular discursive strategy, with parallels with the defensive discourses of established record companies, in the ongoing struggle to retain control over the financial services value chain.

Conclusion

In this chapter we have reflected upon the potential and likely effects of the Internet on constructions, transformations and reproductions of value chains within the fashion, music and retail financial services industries. Despite the relative infancy of e-commerce a number of important points for thinking about the impact of the Internet on the cultural industries can be drawn from the forgoing analysis. Here we want to highlight four points in particular.

First, the commodity form matters. A key factor in explaining the differential impact of the Internet on the music, fashion and retail financial services sectors is the nature of the commodity. Within each sector there have been differing levels of resistance to the digital transformation of the commodity. In contrast to music and money, the materiality of the fashion commodity makes it more resistant to digitalization. Not only does the physical and material qualities of the commodity have important implications for the ability of e-commerce to become significant in each of these industries, but also illustrates some of the limitations of the concept of the cultural industry itself. In the case of the music, fashion and retail financial services sectors, the position of each on the cultural industry continuum (see Power and Scott's prelude to this volume) sheds little light on the likely significance of the Internet for each respective value chain. However, despite the apparent limitations of the cultural industry rubric for thinking about the impact of the Internet on the three sectors in question, one outcome of the emergence and institutionalization of e-commerce in the case of the retail financial services sector seems to have been to amplify the qualitative and quantitative significance of the symbolic content of financial products and services. In this sense the Internet may well have helped to propel the financial services sector along the cultural industries continuum, so that it has taken on many qualities of a "traditional" cultural industry (Lash and Urry, 1994).

However, we are not suggesting that any of these sectors are reducible to a particular reading of the (current) commodity form. Indeed, the second point we wish to make is that although the commodity form matters, our study also illustrates the need to consider the networks of industrial sectors and their topologies in attempting to understand the impact of the Internet. The organizational and spatial impact of the Internet cannot simply be read off from an essentialized account of the commodity form as theorists like Evans and Wurster (1997, 1999) suggest. The Internet's impact upon prototypical cultural industries such as that of music and fashion, or those industries not conventionally defined as "cultural" such as financial services, will be as much determined by the topologies of the various actors and networks which constitute existing value chains, power relations, organizational and discursive politics as by the commodity form. Equally, the potential of Internet technologies to democratize the socially and economically divisive structures of cultural and non-cultural industries alike will depend as much upon the shape and politics of extant and emerging value chains as upon the commodity form.

Third, just as the organizational and sectoral effects of the Internet are much more complex than many technological deterministic accounts would have us believe, so the spatial effects are also more complex than the rhetoric of the "death of distance" (Cairncross, 1997) or the "weightless world" (Coyle, 1997) might suggest (French and Leyshon, 2004). From our analysis of the three sectors studied we suggest that the Internet is reinforcing existing geographies of production, enabling the decentering of consumption, whilst also creating new nodes in the networks of distribution, for example. In addition, it is possible to speculate that the increasing ubiquity of e-commerce might serve to revalorize the event, the display and the performance within contemporary capitalism. As at-a-distance consumption becomes more commonplace, so co-presence may become more highly valued, which has implications for all three industries: fashion shows take on a new significance, as do live musical performances (Leyshon, 2003), while within the retail financial services industry customers are already recognizing that obtaining "face-time" (Rifkin, 2000) with those that administer their accounts and products is becoming more difficult and thus its value is increasing.

Fourth, and finally, we end with some comments on consumption. We suggest that existing accounts of polarized consumption take insufficient account of the role of commodity form, the networks of knowledge and creativity that underpin production in the cultural industries. Simply stated, the form of the commodity, the organizational structures underpinning its production and consumption, and the spaces of its representation all matter fundamentally here, and produce a far more nuanced and culturally-inflected economic landscape than many existing accounts of e-commerce have suggested. The bi-polar narratives that dominate the literature (physical versus virtual, clicks versus bricks, material versus immaterial) seem to be less and less useful in describing and explaining the complex spatialities and temporalities at work when Internet technologies and the cultural industries collide. Rather, we need to be looking more towards convergence within, between and across sectors. Our focus on a range of sectors might offer one way of approaching the array of ways in which industries are cleaved and divided along a number of dimensions: material/immateriality; trust/suspicion or fear; consumption cleavages; and so on. Such an approach, we hope, might in turn give us some theoretical purchase on the ways in which these sectors are evolving in similar and different ways to and from each other.

Notes

1 A brand-based retailer that has left its mark on a range of industries, including record retailing, airlines, railways and mobile telephony.

2 Marks & Spencer is a doyen of the U.K. high street that has traditionally sold high-quality goods to a middle-market clientele, whereas Tesco is the U.K.'s largest grocery retailer.

Bibliography

Bruzzi, S. and Church-Gibson, P. (2000) *Fashion Cultures: Theories, Explanations and Analysis*, Oxford: Berg.

Callon, M. (1991) "Techno-economic networks and irreversibility," in J. Law (ed.) *A Sociology of Monsters: Essays on Power, Technology and Domination*, London: Routledge.

Cairncross, F. (1997) *The Death of Distance*, London: Orion.

Cassidy, J. (2002) *Dot.Con: the greatest story ever sold*, London: Allen Lane.

Connell, J. and Gibson, C. (2002) *Sound Tracks: Popular Music, Identity and Place*, London: Routledge.

Coyle, D. (1997) *The Weightless World*, London: Capstone.

Crewe, L. (2003) "Markets in motion: geographies of retailing and consumption," *Progress in Human Geography*, 27: 352–362.

Crewe, L. and Beaverstock, J. (1998) "Fashioning the city," *Geoforum*, 29: 287–308.

Crewe, L., Gregson, N. and Brooks, K. (2003) "Alternative retail spaces," in A. Leyshon, R. Lee and C. Rand Williams (eds) *Alternative Economic Spaces*, London: Sage.

D'Andrea, G. and Arnold, G. (2003) *Zara*, Cambridge, MA.: Harvard Business School Press.

Dicken, P. (2003) *Global Shift* (4th edition), London: Sage.

Du Gay, P. and Pryke, M. (2002) *Cultural Economy*, London: Sage.

Elson, D. and Pearson, R. (1981) "Nimble fingers make cheap workers," *Feminist Review*, 7 (Spring): 87–107.

Entwistle, J. (2000) *The Fashioned Body*, Oxford: Polity Press.

Evans, P. B. and Wurster, T. S. (1997) "Strategy and the new economics of information," *Harvard Business Review*, 75: 71–82.

—— (1999) *Blown to Bits: How the Economics of Information Transforms Strategy*, Cambridge, MA: Harvard Business School Press.

Feng, H. Y., Froud, J., Johal, S., Haslam, C. and Williams, K. (2001) "A new business model? The capital market and the new economy," *Economy and Society*, 30: 467–503.

French, S. and Leyshon, A. (2003) "City of money?," in M. Boddy (ed.) *Urban Transformation and Urban Goverance*, Bristol: Policy Press.

—— (2004) "The new, new financial system?: towards a conceptualisation of financial reintermediation," *Review of International Political Economy*, 11(2): 263–88.

FSA (2001) *The Regulation of Electronic Money Issuers*, London: Financial Services Authority.

—— (2002) *The Regulation of Electronic Money Issuers: Feedback on CP117*, London: Financial Services Authority.

Gereffi, G. and Korzeniewicz, M. (1994) *Commodity Chains and Global Capitalism*, Westport, CT: Greenwoods.

Gereffi, G. Spener, D. and Bair, J. (2002) *Free Trade and Uneven Development*, Philadelphia: Temple Press.

Gilbert, D. (2000) "Urban outfitting: the city and the spaces of fashion culture," in S. Bruzzi and P. Church-Gibson (eds) *Fashion Cultures*, Oxford: Berg.

Green, N. (2002) "Paris: a historical view," in J. Rath (ed.) *Unravelling the Rag Trade*, Oxford: Berg.

Hughes, A. and Reimer, S. (2004) *Geographies of Commodity Chains*, London: Routledge.

Klein, N. (2000) *No Logo*, London: Flamingo.

Lash, S. and Urry, J. (1994) *Economies of Signs and Space*, London: Sage.

Leyshon, A. (2001) "Time-space (and digital) compression: software formats, musical networks, and the reorganization of the music industry," *Environment and Planning A*, 32: 49–77.

—— (2003) "Scary monsters? Software formats, peer-to-peer networks, and the spectre of the gift," *Environment and Planning D: Society and Space*, 21: 533–58.

Leyshon, A. and Thrift, N. (1999) "Lists come alive: electronic systems of knowledge and the rise of credit-scoring in retail banking," *Economy and Society*, 28: 434–66.

Leyshon, A. Webb, P. French, S. Thrift, N. and Crewe, L. (2004) "On the reproduction of the musical economy after the Internet," *Media, Culture and Society* (forthcoming).

Liebowitz, S. (2002) *Re-Thinking the Network Economy: The True Forces that Drive the Digital Marketplace*, New York: AMACOM.

Lucas, H. C. (2002) *Strategies for Electronic Commerce and the Internet*, Cambridge, MA: MIT Press.

Malmsten, E. Portanger, E. and Drazin, C. (2002) *Boo Hoo: A Dot Com Story*, New York: Random House.

McRobbie, A. (1998) *British Fashion Design: Rag Trade or Image Industry*, London: Routledge.

Nardi, B. A. and O'Day, V. L. (1999) *Informational Ecologies: Using Technology with Heart*, Cambridge, MA: MIT Press.

Phizacklea, A. (1990) *Unpacking the Fashion Industry*, London: Routledge.

Power, D. and Hallencreutz, D. (2002) "Profiting from creativity? The music industry in Stockholm, Sweden and Kingston, Jamaica," *Environment and Planning A*, 34: 1833–54.

Power, D. and Jansson, J. (2004) "The emergence of a post-industrial music economy? Music and ICT synergies in Stockholm, Sweden," *Geoforum* (in press).

Purvis, S. (1996) "The interchangeable roles of the producer, consumer and cultural intermediary: the new pop fashion designer," in J. O'Connor and D. Wynne (eds) *From the Margins to the Centre: Cultural Production and Consumption in the Post-industrial City*, Aldershot: Ashgate Publishing.

Pyke, F. Becattini, G. and Sengenberger, W. (1990) *Industrial Districts and Inter-firm Co-operation in Italy*, Geneva: ILO.

Rath, J. (2002) *Unravelling the Rag Trade: Immigrant Entrepreneurship in Seven World Cities*, Oxford: Berg.

Rifkin, J. (2000) *The Age of Access: How the shift from Ownership to Access is Transforming Capitalism*, London: Penguin.

Thrift, N. (1994) "On the social and cultural determinants of international financial centers: the case of the City of London," in S. Corbridge, R. Martin and N. Thrift (eds) *Money, Power and Space*, Oxford: Blackwell.

Woolgar, S. (2002) *Virtual Society: Technology, Cyberbole, Reality*, Oxford: Oxford University Press.

Wright, M. (1997) "Crossing the factory frontier – gender, power and place in the Mexican Maquiladora", *Antipode*, 29: 278–302.

Zhou, Y. (2002) "New York: caught under the fashion runway," in J. Rath (ed.) *Unravelling the Rag Trade*, Oxford: Berg.

Part III

Creativity, cities and places

5 Creativity, fashion and market behavior

Walter Santagata

Introduction

Creativity-based goods are among the most specialized of all goods. Creativity, like culture, is profoundly rooted both in time and in space. The culture of creativity, or its inherited capital, is inextricably linked to a place, or – in a social sense – to a community and its history. As far as creativity is concerned, time and space matter.

Yet theory on efficient economic behavior is mainly grounded in goods lacking a specific collocation in time or in space. In fact, the more time- and space-specific a commodity becomes, the less efficient is the market mechanism in regulating its production and consumption. The more specialized a good becomes, the less capable is the price system of supplying relevant information, and the less likely is the competition rule to accurately predict results. Thus, the market is an imperfect model for the regulation of creativity-based goods such as fashion, design and art.

This chapter aims to reveal some of the limitations of market behavior analysis concerning creative goods, using the world of fashion as a backdrop for discussion of the economic effects and idiosyncratic characteristics of creative endeavors. In this sense, it contributes to a social interpretation of economic theory, insofar as a society is defined by the place and the time of its development.

The fashion market is apposite for exploring the problems posed by market behavior as it relates to creativity. The enigmatic influence of culture within the fashion industry is manifested in several ways. The culture of creativity – with its recondite fall-out on the originality of an object, its aesthetic and technological quality, and its image – is a distinctive feature of fashion products, whose essential characteristic is that of embodying symbolic values: they are *semaphoric* goods (Barrère and Santagata, 1998; Santagata, 1998a). Moreover, it is the designers themselves, their imagination and fantasy, their views of society and of the history of humankind and their manners and beliefs, that represent the true *deus ex machina* of the workshop-atelier, that mysterious and productive place where fashion is made manifest in the beauty of its forms.

Indeed, the presence of celebrated designers in a given place at a given time is an indicator of a creative environment. The number of creative designers living in Paris during the nineteenth century is impressive (Table 5.1). It also heralds the increasing internationalization of fashion designers, another factor related to the spatial component of creativity.

Table 5.1 Designers in Paris by date of appearance

Prior to *haute couture*	1858	Charles Frédéric Worth and Gustave Boberg
	1871	Jacques Doucet
	1889	Jeanne Paquin
	1898	Les soeurs Callot
	1900	Jeanne Lanvin
	1904	Paul Poiret
Between the two wars	1911	Jean Patou
	1912	Madeleine Vionnet
	1912	Gabrielle "Coco" Chanel
	1919	Edward Molyneux
	1919	Lucien Lelong
	1932	Nina Ricci
	1934	Germaine Barton "Grès"
	1935	Elsa Schiapparelli
	1938	Cristobal Balenciaga Eisaguri
The 1950s	1937	Jacques Fath
	1944	Carmen Mallet "Carven"
	1945	Pierre Balmain
	1947	Christian Dior
	1949	Ted Lapidus
	1950	Louis Féraud
	1952	Hubert de Givenchy
	1953	Pierre Cardin
The 1960s		
	1958	Yves Saint Laurent
	1959	Valentino Garavani
	1960	Karl Lagerfeld
	1961	André Courrèges
	1961	Rosette Met "Torrente"
	1962	Jean Louis Scherrer
	1962	Cacharel (Jean Bousquet)
	1965	Emanuel Ungaro
	1966	Paco Rabane
The 1970s	1970	Jean Charles de Castelbajac
	1970	Issey Miyake
	1970	Kenzo
	1973	Thierry Mugler
	1976	Jean Paul Gaultier
	1976	Christian Lacroix
The contemporary age 1980–2000	Tom Ford, John Galliano, Alexander McQueen, Martin Margiela	

Sources: Grau, 2000; Bergeron, 1998

The felicitous coupling of *haute couture* with *prêt-à-porter* is an excellent example of the creative forces at work in the fashion world. *Prêt-à-porter* and *haute couture* apparel had already been made available separately, but the merging of these two worlds was an absolute Paris original.

Nonetheless, analysis of the economic behavior of the actors – both consumers and producers – is even more revealing of the original, theoretical and social role of creativity and creativity-based goods.

As will be touched on again in the last section of this chapter, consumers have developed a post-modern attitude in their choices, by which they attribute greater value to creative and symbolic factors than to aesthetic and functional characteristics. Consequently, the quest for novelty (Lipovetsky, 1987) – with all of its ramifications within the dynamics of the mimicking manners – is the source of economic behavior affecting social interaction (Simmel, 1904; Bourdieu, 1994; Waquet and Laporte, 1999). The idiosyncratic and inherited character of creativity-based goods, especially in the fashion world, affects economic behavior in two interactive ways: by means of involvement in a community or social group, and by immersion in the productive atmosphere of the cultural industrial districts.

International dissemination of technology has leveled the playing field in international competition, and competition in terms of lower production costs is becoming less and less of a discriminating factor. Consequently, the globalization of markets promotes competition in terms of product creativity. Creativity is the engine of competitive differentiation and success. The amount of creative intellectual property comprising a fashion product overrides the material components by far. Unfortunately, this has led to a burgeoning of the illegal market for counterfeit goods, which is fuelled by the predominance of the intellectual value of a good and by its nature as a public good (Benghozi and Santagata, 2001).

This chapter has three sections. The first presents and discusses three models of creative people and the process of creativity: the creative genius, the manager and problem solving, and creativity as a neurological and social process. The second section is devoted to an economic definition of creativity. The final section deals with the effects of creativity on economic behavior in the fashion world. On the supply side, we will examine the effects of generational waves of creativity. On the demand side, we will examine the high costs of using rationality and their effect on the economic calculus done by consumers of fashion.

Models

Creativity is a hedged and dynamic concept and the search for a complete or absolute definition is an ongoing process. Nonetheless, in the metamorphoses of its rationale we can recognize the tendency for creativity to have become a fundamental resource in post-modern society.

The creative genius

The conventional model of creativity is based on the romantic idea that creativity is the sign of genius, a "superior aptitude of the spirit that makes somebody capable of creations, of inventions which appear extraordinary."[3] (*Le Robert Micro*, 1988). According to this definition, the creative genius is an inspired person. This is the image of creativity as epiphany, a gift received by means of inspiration, meaning "to receive from a mysterious authority, in a way charged of all the characteristic opacity of the creative act, the secrecy of a discovery" (Rouquette, 1973: 10).

This model is especially interested in the narration of the intellectual and psychoanalytical traits of the genius (Kris and Kurz, 1934; Jameson, 1984). Actually, for post-modern culture and particularly for contemporary art, the artist-genius who creates works opposing previous movements and styles is considered to be highly stimulated by psychoanalytic phenomena. Each work of art originates from a hallucination or delirious vision.

This model also explores the whole set of conditions that make it possible to release creativity as a potential property of the spirit. Then it seeks correlations between creativity and a number of human conditions: feelings of guilt, madness, need for autonomy, attitudes towards risk, sex, age, intelligence, money and non-conformism. The image of the creative genius is, therefore, related to a literary and psychoanalytic conception of creativity, as in the case of the creative inventor.

While this model offers a literary description of genius, it would require a lot of jumping through intellectual hoops in order to derive from it a general definition for creativity, and in the process, one would inevitably be forced into the logic of conventional definitions.

Creativity as problem solving

The minimal definition advanced by Herbert Simon is a procedural formula, moving the topic of creativity into a cognitive dimension and anticipating the logic of creativity as a process. Creativity has been defined as a way the mind operates, i.e. "the process with the means of which the mind transforms information into combinations of concepts and produces new ideas" (Goleman, 1997: 18). One may add that creativity is an act of the human brain, manifested as a process which allows us to think and solve our problems – in a way that is commonly considered to be creative (Simon, 1986). Simon's thesis-definition is that creativity consists in good problem-solving.

According to Simon (1986: 1–8), the process that leads to creativity is founded upon three general conditions:

1 *To be prepared.* "Chance, in the words of Pasteur, favors the prepared mind." A casual discovery *per se* does not exist: "It is the surprise, the

departure from the expected, that creates the fruitful accident; and there are no surprises without expectations, nor expectations without knowledge."

2 *To be experts.* Nobody – fashion designers, painters, or musicians – can attain excellence without "an intensive effort to acquiring knowledge and skill about a domain of expertise."
3 *To risk.* Science often requires accepting calculated bets. "Information is only valuable if others do not have it or do not believe it strongly enough to act on it. [. . .] Science is an occupation for gamblers." It is necessary to risk, because, if we want to explore new fields in a creative way, common information is not used instrumental in obtaining differentiated advantages: "scientists require a 'contrarian' streak that gives them the confidence to pit their knowledge and judgment against the common wisdom of their colleagues."

This set of characteristic conditions represents an improvement on prior definitions. Yet, it actually only creates a number of images around a concept. It tells us that fashion designers, for example, with imagination, judgment, taste, intelligence, expertness and a liking for risk are, therefore, creative. It does not reveal the physical sources of creativity. How does the human brain produce creativity? What physical mechanisms of the brain activate a creative mind? When all the secrets of the production of ideas, emotions and feelings have been discovered, we may be able to better define creativity, just as today we know more about the limits of pure rationality following the discovery of the relation between spirit–brain–emotion–social behavior (Damasio, 1994).

The two models so far presented, namely the creative genius and creativity as problem solving, are nonetheless very different. The vision of the creative genius is a mythical concept. From a political and constructivist point of view, this definition fails to assist us in increasing, reproducing and transmitting creativity. How many dressmakers in the fashion world are described with these same words, thus transforming them into extraordinary characters? Nothing could be further from the truth. The procedural approach of Simon and of the contemporary cognitive sciences is instead a significant source of practical suggestions.

Let us now turn to *Descartes' Error* and the neurological theory of creative emotions, as described in Antonio Damasio's remarkable work (1994).

Mind and brain; body and emotions

Body counts, brain counts. Damasio's revolutionary message announces that our whole body is involved in our rational faculty, that the body provides a basic reference to the mental processes. Body and brain play a fundamental role in the faculty of reasoning: their physical function is to process the emotions that the external world sends us all the time. Body and brain, as a unique organism, take part in the interaction with the environment, which is, in turn, partly the product of the human organism's activity.

Emotions count. Emotions are defined as the series of changes which occur in the body and the brain, generally in reaction to particular mental contents. One of the most astonishing discoveries in modern neurobiology has been locating the area in the brain responsible for producing an emotional state in the body. The surprising story of Phineas Gage (1899–1986) tells us of a man who in a labor accident lost the pre-frontal part of his brain, with no apparent physical damage. Further clinical and experimental studies have demonstrated that this part of the brain, the pre-frontal cortex, is responsible for recognition of the emotions, and that, in its absence, we have knowledge without emotions. Patients without emotions still continue to exert an intact and active intellectual faculty, but their decision-making ability is impaired. Damasio notes that the reasoning of individuals lacking emotions proceeds as an infinite sequence of cost–benefit analysis which never leads to a decision. Rationality without emotions proves to be an infinite process. Rationality alone represents the bankruptcy of any process of decision-making. Decision-making is made possible by the presence of what Damasio calls "somatic markers," i.e. images arising from the emotions and which act in the neural structure, allowing the brain to announce that it is necessary to interrupt the reasoning process, which is leading nowhere, and to choose one of the alternatives. The relation between the emotions and reasoning, and the assumption of the existence of somatic markers that facilitate decision-making, give a neurological base to Simon's theory of bounded rationality.

Environment counts. Returning to the theme of creativity and fashion, it is clear that creativity as a problem-solving activity depends on our capacity to interact with a continual flow of emotions. But good emotions influence us positively if we live in a natural or social environment that is rich in such emotions: an environment where there are no intellectual constraints, where incentives and ideas circulate freely and without cost, where freedom to associate ideas and to experiment reaches a climax. As we will see when analyzing the subject of creative management, the theory of emotions is useful in explaining why redesigning a company's organizational and mental environment, so to speak, increases its rate of creativity.

The traditional economic approach maintains that the individual and his mind are a monad, a single entity which simply reacts to a system of price signaling, without any contact or communication with other individuals. In my view, instead, the body and the brain both exist within nature and are submerged in a universe of relations, emotions and interactions. I argue that social interaction and the emotions stemming from them are necessary conditions for promoting problem-solving and creative activity; rationality alone is insufficient. Thus, the productive or research environment are key factors in allowing creative emotions to be released in order to produce, increase and transmit creativity.

Descartes' error was to underestimate the value of the body in relation to the mind: the *res extensa* as opposed to the *res cogitans.* Modern neurological study of the pre-frontal cortex areas reveals that *we are and then we reason.* And our social and natural environment can be modified, just as we can in turn modify our rate of creativity by means of the emotions we experience.

The metamorphoses of our understanding of creativity show a tendency towards a procedural approach. Understanding the origins of creativity, the conditions of its existence and the needs to which it corresponds, is a necessary step if we wish to learn how to produce, increase and transmit creativity. Creativity may best be considered a process characterized by a dual socio-aesthetic and organizational nature.

This process is implicated in every field of human activity, especially in the logic and dynamics of industrial production. The fashion market, in particular, has been deeply influenced by the creative activity of designers and entrepreneur-managers. Creativity in *haute couture* and *prêt-à-porter* apparel (Grumbach, 1993) has existed since the twentieth century. What is new is the development of the concept of creativity, which has developed on two complementary levels: on the one hand, within the subjective sphere of the design of fashion goods; on the other, within the collective sphere of economic organizations and creative management.

Attributes

A brief economic definition

In this section, a brief economic definition of creativity will be given. In the next section, the impact of creativity on market behavior will be examined, using fashion as the market of reference.

Creativity may be considered an economic good produced by the human mind. Creativity is the action that gives rise to something original and unique where nothing was before. This action may take different forms ranging from invention to discovery or even to epiphany. Creativity is the disclosure of novelty.

Table 5.2 lists a wide range of the economic characteristics, in the broad sense, of creativity, classified according to three criteria: the particular nature of the good, the attributes that influence demand and the attributes that influence

Table 5.2 Defining creativity in economic language

Essence of creativity as an economic good	*Characteristics influencing market behavior: demand side*	*Characteristics influencing market behavior: supply side*	
		Products	*Organizations*
Anti-utilitarianism	Symbol and zero information costs	Idiosyncrasy	Generation-based goods
Non-cumulability		Joint production	
Public-ness			
Non-exhaustibility			

supply. First, the basic elements of a minimal economic definition will be outlined. Then, the impact of other attributes on the behavior of the market, in particular in the fashion and clothing sectors, will be explored.

Creativity vs. innovation

In the language of economics, creativity may be conventionally contrasted with innovation. Thus, while creation implies giving life to something that derives from nothing before it; innovation is understood as introducing something new into an existing domain, sequence or process. This conventional view will be explored according to the two main characteristics of creativity: anti-utilitarianism and non-cumulability. In any case, this distinction is rather new within the domain of technological innovation, where creativity is perceived as a usual ingredient of the innovative act, and the focus of analysis is on its Schumpeterian destructive ability or on its being the original source of "disruptive technologies" (Christensen, 1997).

According to the present view, creativity is an essential and autonomous component of human life; basically, it helps develop the intrinsic capacities of the personality. In economic terms, this approach considers creativity to be an anti-utilitarian act, and it stands in opposition to the concept of innovation, which, on the contrary, is registered in the utilitarian system of behavior. Creativity has no purpose; it is an *anti-utilitarian good*. The creative effort produces positive values. It functions as a factor of *self-realization*, it is rich in *intrinsic enjoyment* and in *self-fulfillment*. The assumption that the creator's work is a costly effort becomes less and less valid when one approaches the concept of creative work. In an anti-utilitarian model there is an intrinsic satisfaction in creative work: the more time she/he devotes to this type of work the more she/he is satisfied (Horvath, 1999; Throsby, 2000).

A second characteristic of creativity is that it is a *non-cumulative good*. Creativity is rupture, whereas "normal" innovation as conceived within the frame of a given scientific paradigm (Kuhn 1962, 1977) is a cumulative and incremental process (Santagata 1998b). This feature helps us to more precisely define the anti-utilitarian behavior assumption: the creator offers his working time, because she/he takes pleasure in it. The quality of her/his life does not depend only on consumption, but also on the advisability of choosing to engage in creative work. The "*desire for creativity is one of the most important motivations of human beings in general, and in our post-industrial era in particular*" (Horvath, 1999: 3). This model of behavior, or "art for art's sake property" model, is rather the rule in the creative industries (Caves, 2000). Innovation is instead directed towards the implementation of change (aesthetic, technological or functional). It is a utilitarian good. Innovation is a utilitarian, incremental and cumulative act. It relates to consumption, expressing the objective utility of a product or service. The work required for the process of innovation involves sacrifice and a cost, and implies an external monetary reward.

Other essential attributes

As for the other essential characteristics of creativity, it must be emphasized that the ***intangible*** character of creativity implies that it has to be observed in some material support which contains it and reveals it. The support can be a mere sheet of paper for storing ideas, design and forms; it can also be a more complex object which embodies a creative function. Now, while the support is usually a private good, creativity *per se* and the creativity incorporated in an object is a public good, sharing the features of non-rivalry and non-exclusion. However, just like ideas, creativity must be protected on the market, first of all by establishing laws securing intellectual property. As is well known in the literature on counterfeiting (Benghozi and Santagata, 2001), enforcement of the law is often ineffective, and unlicensed or unlawful producers can copy, at zero cost, any sign of creativity seen, perceived or detected in a creativity-based object. The higher the economic value of the creative and intellectual component of an object, the higher is the incentive to copy.

Finally, creativity is a ***non-exhaustible*** and non-saturable good. The idea or concept serves as an intangible support of creativity. An idea expresses, describes and makes a creative act historical. Unlike natural resources, ideas, resulting from human creativity, are fully exploitable but not exhaustible. The creativity of fashion goods is linked to social evolution and is therefore continuously renewed. Design is linked to its epoch and is therefore always different. Industry enters an inexhaustible field, putting firms on a different footing for confrontation and competition. However, as will be shown in the next section, the evolution of creativity cannot be linear: periods of great creativity and phases of stagnation can always be found, especially in the world of fashion.

Effects

Creativity has different aspects, each of which affects different goods and services: their aesthetic, design, function and productive organization. The most significant cleavage is between the effects on the organization and those relating to the other modes of creativity. The effects on demand are rather concentrated on the aesthetic, design and functional forms, while the effects on supply of an organizational type instead.

Effects on the supply side: the dilemma of a generation-based good and its effects on competition

This part of the chapter is devoted to the effects of creativity on the behavior of agents who are in charge of the supply of creativity-based fashion. The effects of creativity on international competition will be dealt with. Other effects are briefly mentioned: the idiosyncrasy and the specialization of creativity-based products (Santagata, 2002); and the joint production of creativity by both the producer and the consumer (Barrère and Santagata, 1998).

The time/space duality which characterizes the theoretical ground and the dynamics of creativity, shows a significant ancillary trait. Creativity *per se* is the original and specific product of a generation. Now, if the generations in their sequence are affected by various conditions of time and space, the arising dilemma is how to renew the production of creativity while preserving similar traits. Each generation will produce its creative world, but the effects of this phenomenon on the structure of competition are unexpected and significant, in particular in the fashion market. We will see that competition among creativity-based goods is biased by a generational path dependency.

The succession of generations is actually a progressive phenomenon. This is demonstrated by the dynamics of what Bourdieu calls the "field of forces" (Bourdieu, 1971, 1994). As in all fields of cultural production, fashion's field of force is a field of battles: "*avec ses rapports de force physique, économique et surtout symbolique [...] et ses luttes pour la conservation ou la transformation de ces rapports de force*" ("in terms of its physical, economic and above all symbolic relations [...] and of the battles for maintaining or transforming these relations") (Bourdieu, 1994: 140). The coordination of practices and the stakes in the fashion market are, therefore, the texture of a network of conflicts and agreements inherited from former battles. These general conditions affect all generations: the upcoming generations, which try to make space for themselves by opposing the leaders of the dominant *maisons*, as well as the successful ones, which control the official requirements and instances of the fashion world and the production of value.

A new position for a designer can emerge only if the field modifies its structure because the designer must create a new pole in a rather complex process of differentiation. The search for distinction is dominated by the absence of a single principle of cultural justification. The dynamics of the field are endless, implying revisions, arrangements and permanent redefinitions, which are repeated and polarized upon the arrival of each new generation.

Consequently, the rhythm of change in the field of fashion is marked by the succession of the different generations of creators. The reasons can be traced back to the definition of fashion in terms of being an idiosyncratic good, which makes reference to the space/time duality.

In Table 5.1, the evolution of the most renowned Paris *couturiers* is shown by dates of entry in the field and by recurrent waves. The dilemma of reproducing creativity from one generation to the next while conserving the salient characteristic quality is self-evident. Indeed, mother nature distributes talent in unforeseeable ways, with changes in place and time made according to nonlinear trajectories. The atmosphere and the environment in Berlin before the Second World War were not reproducible in the 1950s or post-1989. Historic periods are never reproduced: ideas, culture and manners change, attitudes towards the great social questions fluctuate and styles evolve. As a result, each generation has its own identity, pace and distinction, offering no guarantee of virtuous progress.

An encouraging environment and historical experience seem to show that a critical mass must be reached in order for a wave, or talented cluster, to occur. If this is true, creativity accompanies the lifespan of a generation. Thus, creative waves occurred in French fashion in the 1950s and the 1960s: Christian Dior, Karl Lagerfeld, Hubert de Givency, André Courrèges, Pierre Cardin, Pierre Balmain and, finally, Yves Saint Laurent. Their visibility was strong, although a cohort of epigones always follows a cluster of creativity.

The succession of waves, however, was marked by a *crescendo* of attention, and the major reasons for their emergence can be summarized in three points:

1. An increase in trademark value. Historically, it is apparent that few houses of *haute couture* born in the first half of the twentieth century survived: Mrs Vionnet, Poiret, Worms, the *arbiter elegantiarum* of a glorious age, ceased their activity without leaving heirs. On the contrary, those *maisons* which come into being after the Second World War had more of a chance to survive their founders. It was as though the value of the trademark increased as a result of the increased size of the market. This was especially true in the case of successful fashion accessories. Perfumes, in particular, enjoyed an increasing trademark value. And Chanel, for example, is one of the first houses to survive its creator. Reputation thus becomes an immaterial asset which merits investment.
2. Change in the ownership structure. The ownership of the great *maisons* is characterized by diminished control by the founder's family. The dissemination of ownership shares and access to the stock-exchange market make succession an ordinary routine: there is no longer a financial identification between *couturier*-founder and ownership.
3. Cosmopolite reception. Alongside these arguments, one can stress the presence in Paris of a cosmopolitan tradition of reception. Paris has always been effective in accommodating a wide range of personalities from all over the world, regardless of their name, nationality or social condition. The greatest designers seem to be fearless of changes in country and continent. A cosmopolitan tradition allows intergenerational substitution to occur with neither scandal nor regret. Several lesser-known dressmakers came from abroad: George Vaskène is of Armenian origin, Zyga Pianko is Polish, Gaby Aghion is Egyptian; as well as some of the more renowned ones: Charles Frederic Worth, Christobal Balenciaga, Elsa Schiapparelli, Nina Ricci, Pierre Cardin, Emanuel Ungaro and Karl Lagerfeld. This contrasts with Italy, for instance, where the consumers and even the staff of firms cling to the founders.

A creative wave is not simply the invention of a style, new forms and an original aesthetic. From an economic and organizational standpoint, a new wave brings about the fundamental creation of new processes and products. If simplification is acceptable, one could say that the large French wave of the 1950s and 1960s is highly associated with the invention of the "*haute couture*/modern *prêt-à-porter*" combination; the Italian wave is marked by the organizational flexibility of the industrial districts; and the American wave is characterized by the strategic logic of wide-scale distribution.

But the French creative wave, which revitalized *haute couture* and promoted modern *prêt-à-porter*, shows – like every human phenomenon – a loss of power, a sort of quality depreciation, around the years 1990–2000. A creative wave, indeed, is characterized by a rate of depreciation over time: at its epiphany, creativity attains its maximum level, after which a progressive weakness, which carries the wave to its decline, is perceived. This is a plausible argument for both individuals and their creative drive, and for those organizational patterns in which brilliant pioneers are replaced by epigones – masters in repetition. In Figure 5.1 the wave trajectory has been traced for France, Italy and the U.S., making the time shift quite evident: the Italian wave takes shape approximately twenty years after the French one, and the American wave forms thirty years later than the French one.

The effect of the generational pattern on international market behavior is that the French wave will gain momentum at the precise moment that heralds a faltering in Italian fashion. This is either due to natural or tragic death (Moschino, Versace, Gucci), or to the beginning of the decline of the creative wave. The same advantage will appear with reference to the American competitors, given that the mythical Ralph Lauren is 62 years old. This suggests that international competition in the fashion world is biased by a generational factor, which can also be expressed in terms of path dependency.

These dynamics are confirmed by the history of French fashion. France was the first fashion producer to face the replacement of the greatest founder-*couturiers*. In other words, France was the first country in which a creative generation of dressmakers emerged in the 1950s and 1960s, and it was once again the first country to tackle the problem of the renewal of the great designers in the

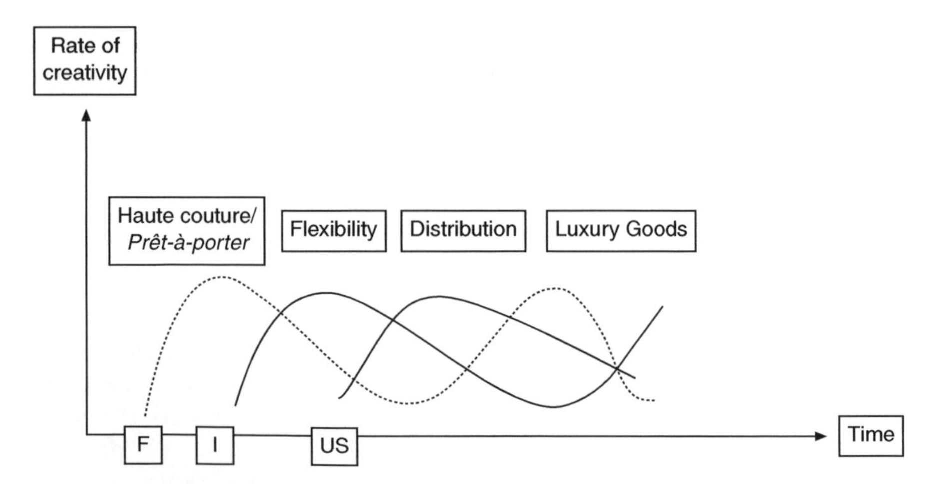

Figure 5.1 The advantage of being the first comer.

1990s–2000: Tarlazzi for Guy Laroche, Ferré for Dior, Montana for Lanvin, Lagerfeld for Chanel, and also Galliano, Ford, Lacroix. Today a "new French wave" is at work. The luxury industries groups are the driving force behind this second French wave. The challenge is, therefore, multiple. At stake in French fashion is the opportunity to seize the positional opportunity related to being the first-comer, thus potentially leading to the development of French superiority in the creative luxury industry. The industry of luxury goods has emerged as the creation of a new product, leader of the French creative wave of the years 2000. It is the new French *creative passion* (Arnault, 2001).

Effects on the demand side

Symbols and creativity-based goods

When creativity is committed to aesthetic values, the form, original functions and goods "created" are laden with symbolic values. This is because the aesthetic, the design, an original function or new forms are recognized by consumers not only for their measurable qualities and quantities, but also because of the signals that touch their heart, soul, emotions, ambition or courage.

For fashion goods, creativity is actually the core of the production value chain. The convention of originality – i.e. the quest for novelty which characterizes this sector's dynamics – implies the formation of a sense of social belonging: people like a particular piece of apparel which is original and allows them to develop a sense of distinction, but at the same time also allows them to develop a sense of social belonging.

The assimilation of creativity-based goods into symbolic goods may take different paths: while symbolic charm can be related to the emergence of originality in the short run, what matters in the long run is a sort of permanent originality, or what might be termed a classic and traditional originality. Creativity allows involvement in a symbolic world in the two ways mentioned above. The first is related to our search for originality and distinction. The second is based on our fidelity to a style that was fully creative at its appearance and which continues to be symbolically representative of a particular status or aesthetic culture.

I will now analyze the effects on consumer behavior, with reference to a large class of goods that can be referred to as *creative symbolic goods*. In these goods are amalgamated both the creative and the symbolic. They cannot be manufactured in the absence of either.

The first characteristic of creative symbolic goods I will discuss is their effect on the economic agents' rationality. In principle, we may say that there is no rationality (*ratio*, *calculus*) without knowledge (*cognitio*). There are emotions, myth, generosity, symbolic adhesion and chance. Without knowledge there can be no economic calculation of the costs and benefit of any action. But the production of knowledge is a costly activity. The system of signs, the languages, the texts, the techniques, the experiments and the information that enable us to gather the necessary data to make a rational choice make up a composite good which has a divergent structure of costs. Paradoxically, as the cost of producing information decreases, the cost of gathering and using information in order to find out if, how, where and when goods and services can be purchased increases.

Our assumption is linked to a conjecture about the increase in cost of use of the market, and, consequently, of economic rationality. In the creative industries (fashion, performing arts, visual arts, industrial design, communication arts including film, TV, the publishing sector and advertising) individuals appear to be more and more attracted by the production and consumption of symbolic goods and beliefs. In particular, they seem to be modifying their choices by replacing complex goods with high information costs, with new goods with low information costs, such as those rich in symbolic values. According to this conjecture, purchase of a specific item does not involve study of its market structure nor evaluation of its hidden quality. Instead, we choose the symbol that allures us and with which we identify. When choosing clothes, we do not calculate the expected costs and advantages of particular items, but the fascination and charisma of the model captivate us.

The assumption of increasing costs of the use of information for rational choices can be examined from at least two points of view. First of all, there is the weakening of the pricing system. In the economic model, prices are necessary and sufficient signals for calculations by the rational consumer. Considering prices as the transmitters of the minimal essential information becomes more and more problematic in light of the quality and the symbolic contents of exchanged goods. Modern goods and services have become more complex, and their qualitative attributes may be covert. The consumer has neither the technical skill nor sufficient knowledge to evaluate them. The cost of rationality is increasing. Secondly, the cost of information gathering and consumption must be considered. In order to gather useful information, the rational consumer must make a time-consuming search, which involves intellectual and physical effort. The consumption of information is an activity whose costs are increasing, even when satisfying decision strategies are available.

Creative symbolic goods: zero information costs

A normal consequence of the increase in the cost of using rationality is that we seek the line of least resistance within the logic of calculation. Among the paths available to us, symbolic values take the place of information concerning goods. Instead of seeking information about the quantitative and qualitative attributes of goods, we entrust – or we are allured by – their symbolic representation. Symbols influence behavior because social actors react to the symbolism they attribute to things. Symbols influence action. Symbols reinforce the common beliefs and the feelings of belonging to a community.

Another interesting point regarding consumer behavior is the characteristic of *zero-degree costs of information*. From an economic standpoint, the reduction of information costs to degree zero is the most significant attribute of symbolic creative goods. It helps explain the emergence of such goods in terms of a micro-economic consumer's response to the increase in the cost of use of rationality.

What matters is the capacity of a good to transmit, at no cost for information to the consumer, a sign which conveys significant information. The consumer is compelled to purchase a given item because the identifying symbolic good conquers him. He need look no further in order to estimate the quantitative and quality contents of goods and services. This behavior corresponds to what Huizinga (1932, Chapter 15) says about the medieval symbolist mentality. The symbol created a "short circuit" in the mind of medieval men and women. Thinking was not systematically the effect of causal connections. A symbol is a thunderbolt, which leaves an imprint on the consciousness of the people.

Conclusions

This short inquiry into creativity and its effects on economic behavior has enabled me to make a certain number of theoretical observations concerning both market demand and supply. Some peculiar traits of creativity have been highlighted: it is an immaterial good which can be produced and transmitted within a positive environment, and the conditions for the production of creativity are dependent on the idiosyncratic nature of creativity-based goods. Moreover, I discussed the idea that creativity is a generation-based good whose major challenge is the continuity of production at constant quality; and that creativity, through its symbolic nature, modifies consumers' choices by providing goods with zero costs of information.

These considerations lead me to be skeptical about the efficiency of the price system in regulating the market for creativity-based goods. Space, time, symbols, culture and the social environment require an economic theory which no longer classifies creativity-based goods as exceptions to be set apart from its main object of study.

The fashion world has been deeply influenced by the emergence of creativity. The behavior of consumers and producers has changed extensively in response to the rhythms and changes of creativity, a good that is both rare and inexhaustible.

Acknowledgments

An earlier version of this chapter was presented at the ACEI 2002 International Conference on Cultural Economics held in Rotterdam. I would like to thank all the participants of the session on creativity for their helpful comments. I would also like to thank three anonymous referees of this book for their valuable comments.

Bibliography

Arnault, B. (2001) *La passion créative*, Paris: Plon.

Barrère, C. and Santagata, W. (1998) "Defining Art. From the Brancusi Trial to the Economics of Artistic Semiotic Goods," *International Journal of Art Management*, 2: 28–38.

Benghozi, P.-J. and Santagata W. (2001) "Market Piracy in the Design-based Industry: Economics and Policy Regulation," *Economie Appliquée*, 3: 73–95.

Bergeron, L. (1998) *Les industries du luxe en France*, Paris: Odile Jacob.

Bourdieu, P. (1971) "Pour une économie des biens symboliques," *L'Année sociologique*, 22: 49–126.

—— (1994) *Raisons Pratiques*, Paris: Seuil.

Caves, R. (2000) *Creative Industries. Contracts between Art and Commerce*, Cambridge, MA: Harvard University Press.

Christensen, C.M. (1997) *The Innovator's Dilemma*, Cambridge, MA: Harvard Business School Press.

Damasio, A.R. (1994) *Decartes' Error*, New York: Putnam's Sons.

Goleman, D. (1997) *Emotional Intelligence*, New York: Bantam Books.

Grau, F.-M. (2000) *La haute couture*, Paris: Puf.

Grumbach, D. (1993) *Histoires de la mode*, Paris: Seuil.

Horvath, S. (1999) "Economic Modelling of Creative Behaviour," *Society and Economy*, 4: 1–11.

Huizinga, J. (1919/1932, French translation) *L'automne du Moyen Age*, Paris: Payot.

Jameson, F. (1984) "Postmodernism, or the Cultural Logic of Late Capitalism," *New Left Review*, 146: 53–92.

Kris, E. and Kurz, O. (1934/1979) *Legend, Myth, and Magic in the Image of the Artist. A Historical Experiment*, New Haven, CT: Yale University Press.

Kuhn, T.S. (1962) *The Structure of Scientific Revolutions*, Chicago: Chicago University Press.

Le Robert Micro (1988) Paris: Dictionnaries Le Robert.

—— (1977) *The Essential Tension*, Chicago: Chicago University Press.

Lipovetsky, G. (1987) *L'empire de l'éphémère*, Paris: Gallimard.

Santagata, W. (1998a) *Simbolo e merce*, Bologna: Il Mulino.

—— (1998b) "Propriété intellectuelle, biens culturels et connaissance non cumulative," *Réseaux*, 88–9: 65–75.

—— (2002) "Cultural Districts, Property Rights and Sustainable Economic Growth," *International Journal of Urban and Regional Research*, 26: 181–204.

Simmel, G. (1904) 1957 Fashion, *American Journal of Sociology*, 62: 61–76.

Simon, H.A. (1986) "How Managers Express their Creativity," *Across the Board*, 23: 1–8.

Throsby, D. (2000) *Economics and Culture*, Cambridge: Cambridge University Press.

Waquet, D. and Laporte, M. (1999) *La mode*, Paris: Puf.

6 The designer in the city and the city in the designer

Norma Rantisi

> What is seen is fashion . . . what is difficult to see is the engine that drives fashion, the industrial organization that produces it. And yet this is one of the most interesting and original aspects of fashion, an aspect that links up with the concrete plane that lies beyond its apparent frivolity and insubstantiability.
>
> (G. Malossi, *The Style Engine*)

Introduction

Studies on the origins of style and creativity are no longer the domain of art historians and cultural critics. The competitive pressures of a global economy and a growing segmentation of mass markets have made aesthetic innovation the new mantra of late capitalism and, consequently, a new focus in the fields of management, business and economic geography (see, for example, Bjorkegren, 1996; Leadbeater, 1999; Scott, 2001; Howkins, 2001; Florida, 2002). These new studies are centered on the increasing convergence between culture and the economy, as the "symbolic" attributes of goods and services are now deemed key elements of productive strategy. The competitive nature of such goods and services (or "cultural products"), it is argued, flows from their ability to entertain, provide a form of social identity, or confer status, over and above their utilitarian value (Harvey, 1989; Lash and Urry, 1994; Scott, 1996). And the key questions that this line of inquiry poses are how best to enhance the creative process and to realize its economic value.

Conventional analyses have privileged the individual artist as the primary source of aesthetic innovation and have tended to focus on psychological processes (e.g. Koestler, 1990; Gardner, 1993; Ghiselin, 1996). Such an emphasis on an autonomous *créateur*, however, precludes an examination of the social and economic context in which the individual operates and the opportunities or constraints that such a context can afford. The creative process does not occur in a vacuum. And the same economic imperatives that have made aesthetic innovation all the more significant have also made the process all the more challenging for an individual to undertake alone.

Accordingly, a focus on the individual has given way to a focus on the broader set of relations that are implicated in the production of culture. First articulated by Hirsch (1972), this latter perspective assumes that cultural products are the outcome of a *series* of interrelated processes, encompassing creation, production, distribution and consumption. Particular attention is drawn to the role of intermediary organizations – such as wholesalers or media outlets – in establishing the links between creators and consumers. As marketing and distribution channels, intermediary organizations serve as gatekeepers that identify and promote particular categories of goods and services, and, together with the creators and consumers, are said to constitute *cultural industries* (Hirsch, 1972, 2000).

More recent extensions of this "industry" approach, most notably the works of Allen Scott, have also highlighted a *spatial* dimension to the process of cultural production. According to Scott (1997, 2000), cultural industries tend to agglomerate in certain places to more easily access specialized labor, supply firms and support services, and to tap into the creative energies of other industries. And due to their diversity and concentration of activities, cities are particularly well suited to support these economic functions. Over time, some cities may even develop lines of specializations that create symbolic associations between the place and the cultural product, providing the product with a competitive advantage. These associations can also serve to draw a larger number of support firms and organizations to the city. In this way, the product and its creator/s reproduce the city, and the city (as a set of institutions and images) becomes a critical input into the production process.[1]

New York fashion as cultural industry

The objective of this chapter is to examine this recursive relation between the city and cultural production through the prism of fashion. As a commodity which is valued for its social and symbolic attributes relative to its utilitarian ones, fashion by definition implies continuous innovation. The creator – in this context, the designer – must always keep in lockstep with shifts in consumer preferences (Agins, 1999a). An industry system constituted by a host of intermediary institutions assists the designer in realizing this objective. As suggested by Scott (2000), these organizations tend to concentrate in urban settings, a trend that has been highlighted in numerous studies (Helfgott, 1959; Crewe, 1996; Crewe and Beaverstock, 1998; McRobbie, 1998; Scott, 2002).

Within the U.S. context, the undeniable center for fashion is New York City. It is home to world-renowned designers such as Donna Karan, Ralph Lauren and Calvin Klein. It is the hub for the top ten fashion magazines and the top two fashion design schools in the country. And, alongside Paris and London, New York is host to one of the largest international fashion shows. The city's industry accounts for approximately 20 percent of the value added for U.S. women's wear and generates seventeen billion dollars in manufacturing and wholesale output (U.S. Bureau of the Census, 1997; New York Industrial Retention Network, 2001). While Los Angeles has recently surpassed New York City as the national center for apparel production, it lacks the design, marketing and retail infrastructure that characterizes the New York fashion economy (see Scott, 2002 and Rantisi, 2004). Moreover, the New York industry has a specialization in women's wear, which accounts for over 75 percent of industry employment, and a dominance in the higher-end segments (e.g. dresses) (U.S. Bureau of Labor Statistics, 2001).

Within New York, almost two-thirds of industry is based in the historic Garment District, a four-by-six block area in the western half of midtown Manhattan, dubbed "Seventh Avenue" by industry insiders. This district is comprised of a dense concentration of apparel manufacturers and contractors, as well as a wide range of supply and support services. Such services include design schools, forecasting services, textile design studios, fabric suppliers, trade shows, fashion magazines, fashion shows and buying offices. The Garment District is also located in close proximity to major retail centers, such as Fifth Avenue and Soho, and to major cultural institutions in the city, such as Broadway and the Metropolitan Museum of Art, as well as popular nightclubs, cafes and restaurants. Together, these cultural institutions and apparel-related services (i.e. "cultural intermediaries") constitute a local industry system on which designers can draw, making New York an illustrative case in which to explore the city–fashion nexus.

In examining New York fashion as a cultural industry, I focus on the high-end segment of women's wear.[2] The high-end segment, in contrast to the moderate-to-low end segment, relies more on aesthetic innovation than imitation as a

Plate 6.1 Donna Karan advertisement at corner of Houston and Broadway, Manhattan, New York (Source: Photograph taken by Noah Najarian, reproduced with permission).

competitive strategy,[3] and for apparel, the "moment" of aesthetic innovation is design. Following the lead of Hirsch (1972, 2000) and other cultural industry proponents, I look at how cultural intermediaries contribute to the *realization* of a design by supporting its materialization and commercialization. However, this analysis goes beyond past studies by examining how intermediary institutions in the city can also act as a "creative field", offering aesthetic stimuli that shape how a design is *conceived* (Scott, 2000). I consider the role that the city can play as a source of both "art" and "commerce" within the fashion production process.[4]

The city as "art"

The first step in producing a cultural product such as fashion is deciding *what* to produce. The decision of what to produce entails the selection of fabrics, colors and silhouettes for a given season's collection, and the process generally begins with a concept that can be inspired by a range of influences. Interviews with high-end designers reveal that some of the most commonly employed sources of inspiration for that concept include New York-specific cultural institutions, which could be broadly classified into two sets: complementary cultural institutions (i.e., other design-oriented fields) and fashion-related cultural institutions, (e.g., those fields directly tied to the fashion industry). The following sections analyze how these sets of institutions provide designers with creative stimuli that can encourage aesthetic experimentation.

The role of complementary cultural institutions

Apart from being an international fashion capital, New York is a renowned center for the performing and visual arts. Prior to the mid-twentieth century, New York was said to draw its artistic inspiration from Europe, with New York artists often visiting European capitals, particularly Paris, to acquire insights into new avant-garde techniques. In the aftermath of World War II, however, the power of balance in the art world shifted from Paris to New York, as many renowned European artists immigrated to the States. The war and the subsequent closure of Paris afforded New York talent the opportunity to acquire the training and resources needed to sustain their own industry and to develop a uniquely modernist American style, as exemplified in the works of Georgia O'Keefe, Miles Davis and George Moses (Guilbaut, 1983; Scott and Rutkoff, 2001). While the American style has evolved over time, much of the nation's artistic achievements remain centered in New York. Local talent is supported by the presence of notable cultural institutions, such as the Metropolitan Museum of Art, the Guggenheim Museum, Carnegie Hall and the Lincoln Center for Performing Arts, which reinforce New York's position within the international cultural circuit.

Indeed, the city's status as a cultural capital is an asset not lost on fashion designers. When asked about their sources of inspiration, over one-third of the designers interviewed cited architecture, art exhibits, opera and theatre as primary sources. The discussion of these sources centered on the conceptual elements that such industries could contribute towards the visualization of a design. According to one high-end designer, who cited New York architecture as a key influence, "architecture has always been the avant garde of the design industries . . . it plays an important role in defining the forms, textures, colors for a given period" (personal interview, 2000). According to another designer, "designing comes from my head. Not that that doesn't incorporate everything around you. I mean if you go to a great opera or you see a great painting, there's a color stimulus (there)" (personal interview, 2001). The relation between the aesthetic qualities of New York and a fashion designer's vision was also highlighted in an *AsianWeek* interview with New York-based designer Vivienne Tam. Citing the New York skyline as a point of reference for her Spring 2001 collection, Tam states, "Many of the prints and patterns in the collection are the result of the views from my terrace . . . I love watching the light shimmering as it plays with the architectural corners and angles of buildings against a grey and bluish evening sky" (quote taken from Harlan, 2000, www.asianweek.com/2000_09_28/feature.html).

The link between theatre and fashion has also been strong in New York City, with the grandeur and detail of costumes contributing to the spectacle of a theatrical production and providing a source of inspiration for future designs, (Owen 1987; Museum of the City of New York, 2000; personal interview with a high-end designer, 2001). Historically, the bond between these two artistic fields was formalized in the New York context with the establishment of the Costume Institute (formerly, the Museum of Costume Art) in the late 1930s. The Institute was initially founded by members of a local theatre group (Chase and Chase, 1954). Later, with financial backing from fashion industry elites, it became part of the Metropolitan Museum of New York, where it exists today as a separate curatorial department.

The Costume Institute serves as a valuable resource for designers not only because it houses a rich collection of costumes that spans 300 years and sponsors special exhibits but also because it has become a repository for old design sketches and patterns passed on by industry elites, which contemporary designers can now access. More recently, the Fashion Institute of Technology (F.I.T.), one of the leading design schools in the country, has also built a collection of old design sketches and patterns, often acquired from past alumni. One high-end designer, with an old design sketch hanging on her wall, acknowledged, "while most designers won't admit it, many of them rely on old design sketches for ideas . . . famous designers are often spotted at the Costume Institute and F.I.T. library" (personal interview 2000). Indeed, this trend was confirmed by a Costume Institute librarian, and during my own visit to the Institute, the design assistant of an established Garment District designer was sitting across the table browsing through an album of old sketches.

In addition to the more formalized institutional fields, local consumers constitute another cultural institution that shapes the conceptualization of a design. Over one-half of the designers interviewed cited people on the streets or in local nightclubs and parties as key sources of influence. As one designer contends, "a lot of times, it's what you see everyday that helps you stay fresh. Everybody [here] is so fashionable. Stand outside for a minute and you can see so many different looks' (personal interview, 1999). In describing how local fashion is appropriated, another designer explains:

> People inspire me. Watching people on the street . . . and the way they put it together. You know, I saw a girl who had on a dress yesterday . . . I said "that dress was so light." But then I looked down, she had on these heavy, heavy tights. Then, I said "that makes a lot of sense; somebody was thinking." Then, I saw another woman. She had on an olive green leather coat but then she had a light green cashmere shawl draped around her. And I just thought that was so chic, it was really good, because I'd never liked leather.
>
> (Personal interview, 2000)

An ability to draw on the social fabric of the city is enhanced by a concentration of diverse lifestyles. "Any feeling you want to get, you can go get. A sense of China, go to Chinatown, or Little Italy, all different cultures in one direct centre within a couple of miles of each other, feet of each other at times," says one designer in explaining what underpins the city's artistic pulse (personal interview, 1999). Within this context, New Yorkers themselves represent living cultural artifacts that are deemed important markers of cutting-edge styles and savvy consumers of the culture that they help to produce.

The role of fashion-related cultural institutions

Another set of institutions that serve as a source of design inspiration are those institutions that cater directly to the industry and provide a space in which to observe and monitor alternative approaches to fashion. The most obvious example of such fashion-related institutions is the New York retail market. Since the New York women's wear industry was initially developed in the mid-to-late 1800s by immigrant wholesalers and retailers, it is not surprising that today the city hosts a large concentration of retail shops, ranging from department stores and specialty stores to independent boutiques. The primary retail district in the city, which can be dated back to the early 1900s, is situated on the east side of Manhattan, along Fifth and Madison Avenues (Jackson, 1995). These shopping corridors (otherwise known as "Fifth Avenue" and "Madison Avenue") include such established institutions as Bloomingdales, Saks Fifth Avenue and Henri Bendel, and represent one of the most vibrant shopping centers in the country. They comprise the high-end segment of the local market and are said to cater to the "corporate" woman.

Within the last twenty-five years, other distinct retail districts have emerged throughout the city, particularly in downtown Manhattan. The Soho district, for example, has become a shopping destination for residents and tourists in

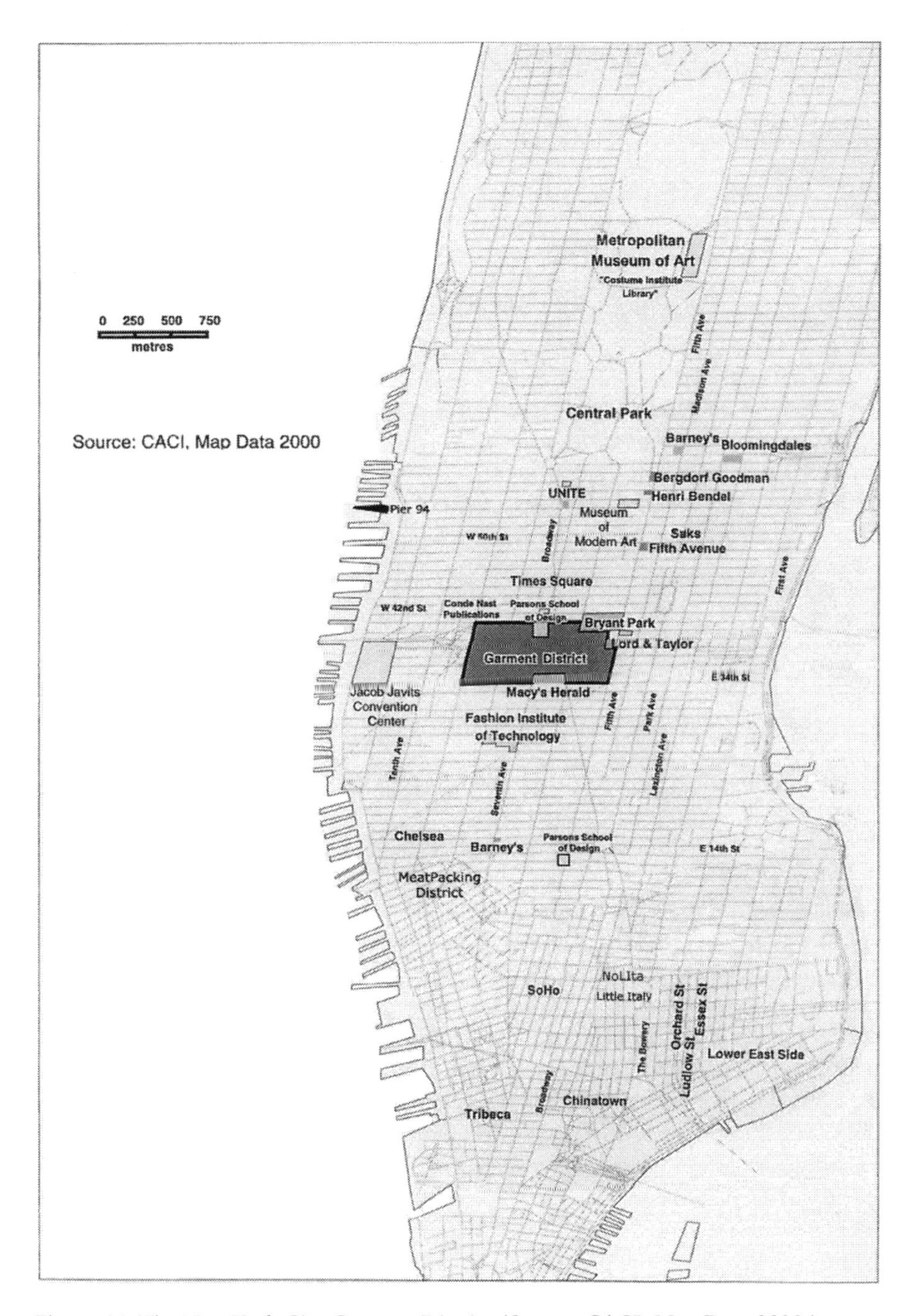

Figure 6.1 The New York City Garment District (Source: CACI, Map Data 2000.)

search of more alternative, avant-garde styles. It was formerly home to a large number of independent designers who would sell their clothes in their own boutiques. More recently, the area has become gentrified, as more established corporate institutions, such as J. Crew and Banana Republic have been opened stores there. Consequently, a number of independent retailers and designers have now set up shops in adjacent areas to the east of Soho, creating new bohemian-style enclaves. These areas are commonly referred to as Nolita ("North of Little Italy") and Orchard Street, and they have become shopping districts in their own right, as evidenced by their mention in guide books and news articles (e.g. Pratt, 1997; Holusha, 2002). In contrast to the department and specialty stores to the north, the shops in these districts offer more exclusive and "edgier" products since they are not producing for a mass market (Rantisi, 2002a).

Since each of the shopping districts in the city has a distinct character, the retail market as a whole offers a diverse array of stimuli. Designers can look to the creations offered by their direct competitors or to designs that are offered up at different price points and for different product markets (Rantisi 2002b). Indeed, many of the designers interviewed said that they would regularly visit the stores to keep abreast of what is "trending" on the retail scene. With regards to the merits of shopping the local market, one sportswear designer had the following to say:

> When I go to Saks; it helps to inspire me. It's like I have some frames to work in . . . I can get a design off a wedding dress and make a sweatshirt out of it. Just because we are doing sportswear, I don't only look into sportswear. Because you can get different ideas from different things . . . even if it's just the way they did a certain stitch or fabric or they way they combine fabrics. It could be something small, not the entire garment, but the smallest thing on that garment.
>
> (Personal interview, 2000)

The high-end designers interviewed were careful, however, to draw distinctions between getting ideas from the local market and copying them. They see themselves engaged in a creative process that borrows elements from existing designs but then reinvents them with the development of new products.

Another fashion-related institution that serves as a source of creative stimuli is the bi-annual New York fashion week. In 1993, the Council of Fashion Designers of America (CFDA) established an organization that took on the task of coordinating and centralizing the fashion shows in the city by hosting them over the course of a week in the tents at Bryant Park, a park located to the north-east of the Garment District (see Figure 6.1). Appropriately titled "7th on 6th,"[5] one objective of this organization has been to promote the exposure of local design talent to international buyers and the worldwide press, in an effort to put the New York industry on par with that of London, Paris and Milan (personal interview with the Executive Director of "7th on 6th," 2000).

And the growth and success of the shows is evidenced by the acquisition of "7th on 6th" in 2001 by IMG, a global marketing company (Chen, 1999).

For high-end fashion designers, the centralized showing of the collections provides another means for observing the creations of their competitors. In most cases, the collections presented on the runways are not the collections that will actually get manufactured. The clothes that are presented are viewed more as theatrical, since the primary objective of the runway is to generate excitement and spark an interest in the designer (Chen, 1999). However, the runway styles do capture the general concepts for the collections that are to be sold – i.e. a certain color, stitch or trim that will be used – and, in this way, they can have a "trickle-down" effect in the industry by defining the themes for a given fashion season. Several of the designers interviewed indicated that the fashion shows shape industry trends because they are "directional," "inspirational," and "fashion forward" (personal interviews with high-end designers, 1999, 2000).

Other sources of design inspiration

In addition to the complementary and fashion-related cultural institutions highlighted above, the high-end designers interviewed cited a number of non-local sources of design inspiration. Internationally renowned designers, such as Yohji Yamamoto or Miuccia Prada, for example, were a primary source mentioned. Other non-local influences cited include travels abroad to Europe, India and Africa, and old movies. All of the respondents cited more than one influence; however, the vast majority of the responses provided (over 80 percent) were specific to the New York cultural economy. Such findings clearly reveal a strong and positive relation between the city and the creative process.

The city as "commerce"[6]

Cultural institutions within the city do not only serve to encourage innovative dynamism by exposing designers to distinct ways of seeing and doing fashion; they also play a critical role in translating the designer's vision into a tradable commodity. Fashion-related institutions in particular are significant intermediaries in the process of linking designers to consumers. Historically, the emergence and concentration of these key institutions in the New York Garment District occurred in the earlier part of the twentieth century, at a time when the city was undergoing rapid urbanization. The high-volume production of ready-to-wear apparel, in which the city specialized, demanded ancillary manufacturing, marketing and distribution services to facilitate the coordination of a mass-market enterprise (Scranton, 1998; Rantisi, 2004). Over time, these services have come to function as part of a well-established production system, characterized by shared conventions and business practices. They influence the process of cultural production today by providing designers with market-sensitive information, the skills for using that information in the design of products and the mediums by which to market those products. In the sections that follow, I examine how this occurs by looking at the role of the city in the materialization and the commercialization of fashion, respectively.

The role of the city in the materialization of the design

The materialization of a design refers to the process of giving material form to a concept, deciding its color, shape and texture. The institution that provides the foundation for such a process in the case of high-end designers is clearly the fashion-design school. Ninety percent of high-end designers interviewed indicated that they received their initial training from either the Fashion Institute of Technology (F.I.T.) or Parsons School of Design, both located in or near the Garment District. Parsons is the older of the two schools, established in 1896. Initially, the school was oriented towards the fine arts and the curriculum focused on painting and drawing, but it became more oriented to the applied side of design in subsequent decades in response to industrialization and the growth of mass-produced consumer products. F.I.T. was established in 1944 by local industry elites, many of whom were immigrants and were concerned with the training and education of a new cadre of apparel manufactures to serve an expanding industry in the post-war period. Since its inception, the school has been oriented toward training students for careers in the industry (personal interview with the Chair of the Fashion Design program, 1999).

How do these training institutions help designers balance creativity with the demands of commerce? First, they shape the way designers "read" the city. They promote a strategy of viewing the city as "art" by sponsoring field trips to museums and art galleries and by providing reduced tickets for the operas, shows and theatres. According to the Chair of Fashion Design at one of the schools, students are also encouraged to walk around the neighborhoods and to take in their diversity, for example their restaurants, retail shops and architecture, since "in New York City, just walking through the streets – Soho, the Village – is an experience" (personal interview, 1999). This exercise sensitizes students to the broad range of stimuli that a city can offer. Moreover, it helps them to identify trends that cut across diverse cultural institutions and can be deployed in the establishment of design concepts that are innovative, yet fashionable (i.e. acceptable) within the broader cultural economy.

The second way in which the schools contribute to the development of fashion is through their explicit merchandising orientation. In addition to traditional courses in art history and draping, the fashion-design programs at F.I.T. and Parsons cover business issues (e.g. apparel manufacturing, the costing of resources, new trends in fabric development) in their apparel design courses or offer separate courses in business and management. Both programs also have strong ties to the local industry, and industry leaders, who are often alumni of the programs, play an important role in the schools' curricula by serving as guest lecturers, teaching senior-level seminars and judging final-year design projects. Internships are another means by which industry links are forged. They help students to establish key networks and acquire first-hand experience of the realities and constraints of getting a product to the market before formally entering the industry (personal interviews with the Internship Director for F.I.T. and the Associate Chair of the Fashion Design Program at Parsons, 1999).

Many of the high-end designers interviewed acknowledged that the career-oriented nature of the local design programs have facilitated their transition into the workplace. The programs encourage them to be creative with their concepts while at the same time training them to channel their creativity for commercial purposes. The success of this orientation is evidenced by the fact that, at F.I.T., over 85 percent of students find employment in the local industry upon graduation, approximately half staying on at the companies where they held internships (personal interview with internship director at F.I.T., 1999). Moreover, the ability of graduates and interns to perform their required tasks was confirmed by several of the designers who hired them on as assistants. According to a New York-based European designer, who compares her own training process in Europe with those of the local interns she has hired:

> American fashion, the way its taught, is all about merchandising. It starts from the concept of merchanding. It does not start from an idea. European schools are more about the idea ... here, its about business, its all very worked out, the [fashion] seasons, the weight of the fabric, the price points. And my interns here, the way they teach them, they know all this stuff. I was never taught it in school.
>
> (Personal interview, 2000)

In addition to the design schools, a number of fashion-related institutions in the city assist designers in realizing (or "materializing") their vision. In particular, forecasting services and trade publications serve as important sources of market information that designers consult when deciding on specific design elements (e.g. the colors, fabrics and styles) that they will employ. Forecasting services, for example, will undertake market research and provide designers with the trends eighteen months in advance, so they can anticipate what materials or resources they will need to acquire and, if needed, validate expenses to upper-level management. According to the director of one forecasting service, her office acquires information through "endless, constant research ... [by] shopping stores, museums, galleries, movies, and fashion shows to keep a pulse on fashion trends and popular culture" (personal interview, 2000).[7] High-end designers typically employ these services when they are moving into areas of specialization that are outside their traditional design focus or silhouette or when they are starting a new project, since the market research can be too costly or time-consuming for them to undertake alone. Approximately 30 percent of the high-end designers who were surveyed indicated that they acquired trend reports or individualized consulting from local forecasting offices. However, not all designers can afford the services, which can cost up to $5,000 annually, and some designers admitted to acquiring them in indirect ways by attending free seminar presentations or obtaining free sample reports (personal interviews, 1999, 2000).

Trade journals serve a similar role as forecasting services in assisting designers to navigate the market and to identify the appropriate trends for their price points and product specializations. Such publications will often feature best practices in the industry, highlighting new developments in fabrics and who offers them or new design, marketing and production strategies. *WWD* (formerly *Women's Wear Daily)* is the most popular of the New York-based publications. In existence for over ninety years, this daily journal serves as the primary source of business news for apparel manufacturers, retailers, suppliers, fashion educators and other fashion-related organizations, and has a circulation of 55,000. All of the designers and key industry actors interviewed said they would regularly consult *WWD*. According to one designer, *WWD* is the most significant resource because "it tells you the market, the newest trends, and it will tell you who's doing well, what the stores are that you are competing with . . . you know, the business of the day" (personal interview, 2000).

Forecasting services and trade publications are particularly relevant for designers at the stage in which they are making their samples. Once samples are made, the key institution shaping the subsequent production runs is the retailer. Retailers can influence the production process through their decisions on which samples they will purchase, the quantity to be purchased and the price that they will pay. These choices affect the look and the quality of the garment because they determine the kinds of materials a designer can obtain and the manufacturing options they will have. In addition, retailers may make specific recommendations as to the colors or fabrics that a designer should use for a particular garment, based on their knowledge of last season's sales or of the sales of competing lines. And the leverage that retailers have over the design process has been increasing over the past two decades due to a concentration of retail buying power in the hands of a few large companies.[8] The retailers will not dictate the designs *per se*, particularly since high-end designers can offer them quality products that are distinctive in the marketplace, but they establish the parameters within which designers create (Lubow, 1999). One designer explains how their sales input helps her to establish a map for what is needed "instead of just designing twenty dresses because I love dresses, when we are not going to need twenty dresses because we are just going to sell one" (personal interview, 2000). Thus, the retailer within the local fashion production system ensures that art works in the interest of business.

The role of the city in the commercialization of the design

The transformation of a design into a material object is an essential part of the fashion-production process, but for the object to constitute a commodity, it must formally enter the marketplace as a good to be bought and sold. Within the city, there are a number of cultural intermediaries that regulate this entry. Often referred to as "gatekeepers" (Hirsch, 1972), the primary role of these institutions is to link the designer to the consumer by serving as distribution or marketing channels. Their role, however, is not merely that of regulator. As Zukin (1991) suggests, these arbiters of taste represent a "critical infrastructure" that interprets the cultural product for a consumer and, in the process, may alter its symbolic value.

In the case of fashion, two key institutions for marketing – and producing – culture are the fashion show and the fashion magazine. The fashion show represents one of the few occasions in which designers can control how their concepts are exhibited. As highlighted above, the objective of this event is to "turn heads" rather than display wearable or affordable clothes; thus designers will often use the most luxurious fabrics available on the market, employ celebrity models and, in some cases, even hire a live band to perform on the runway – all at a cost of thirty to fifty thousand dollars per show (personal interviews, 1999, 2000). Cultural institutions and artifacts in the city figure centrally in this effort to cultivate symbolic attributes and associations. For example, to reinforce the company's association with urban hip-hop, one designer used the following approach: "at the end of the show, we had people come out and beat on the drums, actually garbage cans, [like] 'bring in the funk' which was an idea taken from an urban musical playing on Broadway" (personal interview, 2000). In this way, the show highlights the entertainment value of fashion by creating a spectacle around the designer's collection (personal interview with the Executive Director of "7th on 6th," 2000).

The development of "7th on 6th," as described above, has elevated the significance of the shows as a marketing resource by coordinating the locations and times to ensure maximum visibility for each participant. Since this event is not open to all designers, participants also benefit from the prestige of having been selected by a committee of their peers. Indeed, several participants acknowledged the importance of the event in helping them to get press coverage or establish new retail accounts. According to one designer, "We got a lot of publicity from '7th on 6th.' There were a lot of magazines that didn't know of us before, that were mentioning us" (personal interview, 2000). Through their coverage in magazines and in TV/newspaper fashion segments, the shows also serve to generate a broader interest in fashion among consumers. As one fashion consultant explains, this wide publicity is the event's *raison d'être* because "in other countries . . . they're always doing fashion, have major fashion editorials in their newspapers. [But] that doesn't happen as often in America. So, it is important to get the customer excited twice a year" (personal interview, 2000).

Interpreting the runway shows is one means by which fashion magazines contribute to the commercialization of fashion. The magazines also play a "gatekeeper" role by providing editorials on the styles offered in the stores and by providing space for designers to advertise their products. Since New York City is a hub for the publishing industry in the U.S., established fashion magazines – including *Vogue* and *Harper's Bazaar* – with a nation-wide circulation of thirty million are based in the city (Fashion Center BID, 2001). This location helps magazine editors to keep abreast of fashion trends since they can easily visit the designers' showrooms and attend industry-related events. It also gives designers opportunities to network with editors, since they often go to the same restaurants, clubs or cultural events (personal interviews with high-end designers, 1999, 2000). Both the magazine editors and designers concede that these social ties are significant for acquiring visibility and status within the industry. According to an assistant editor at one of the established magazines, a number of the designers who received favorable magazine editorials would often drop by the office to visit the editor-in-chief (personal interview, 2000).

While fashion shows and fashion magazines are directed primarily at the consumer, key trade publications, trade shows and buying offices serve as intermediaries that are directed primarily at the *retail buyer*, the individual who ultimately decides what products will appear in the stores. The potential significance of the trade publications, such as *WWD,* has already been highlighted above. However, there is one trade publication, *The Tobé Report*, that is targeted exclusively to buyers and deserves special mention. Established in 1927 by a leading retail consultant, this weekly report features upcoming styles and evaluates specific apparel designers to assist buyers in deciding which items to purchase and where to purchase them. In contrast to the fashion magazines which cover trends for the current season, the value of this report is that it follows the industry season, which is six months ahead. Many of the retailers interviewed said that they would consult *The Tobé Report* when making decisions about what products to purchase, and the President of *Tobé* has said that almost all major retailers in the country are among her clientele (personal interview, 2000).

The apparel trade show is another marketing institution that has become popular in recent years. Traditionally, a retail buyer would go from one designer's showroom to the next to view the range of products available on the market, and though the showrooms are concentrated along major streets in the Garment District, these individual visits would command a lot of time. Many buyers will now attend trade shows that are held at the major convention centers, such as Pier 94 and the Jacob Javits Center. According to one Garment District sales representative, the popularity of these shows can be attributed to the fact that "they bring hundreds of apparel manufacturers together under one roof, making it easier for buyers to shop and compare local offerings" (personal interview 1999). Buyers also prefer them to the fashion show because of the difficulties of viewing products displayed on the runway (Agins, 1999b). Although designers are generally restricted to small cubicles at these shows, this marketing option has become increasingly appealing with the introduction of juried shows. The "Fashion Coterie," sponsored by ENK International, is one example where the participants are selected by a jury of industry elites (personal interview with high-end designer, 1999).

A final institution that acts as cultural intermediary is the resident buying office. Dating back to the early-to-mid-1900s, these offices were created to link out-of-town buyers with local manufacturers at a time when the New York market was expanding. Today, they edit the market by identifying those designers that will best suit the needs of a particular buyer's clientele. In explaining the process, an account executive at one of the largest buying offices maintains, "for any given fashion season, we will view all the designers. We will identify the latest trends, and from an aesthetic vantage point, determine which manufacturers best interpret the latest trends. We also ascertain from the store what their mix and their specific needs are" (personal interview, 1999). Most of the buying offices specialize in particular price points. Gregor Simmons, for example, will cater only to specialty stores dealing with high-end products (personal interview with an account executive, 1999). This specialization enables the offices to develop a better command of their respective markets and, thereby, enhance their standing as authorities or "arbiters" within the local production system.

Situating the "designer" in the city: some concluding remarks on the city–fashion nexus

An examination of the city as a source of both "art" and "commerce" reveals how a number of intermediaries assist the designer in the process of producing fashion. On the one hand, these institutions serve to inspire by presenting designers with cultural artifacts, which they can exploit in the development of their own creations. The institutions, however, also play a directive role in guiding the design innovation process. By rendering what is possible, affordable and acceptable in the realms of production and consumption, they shape how a design is conceived, manufactured and marketed. The city, as such, establishes the social and economic parameters of creativity, giving meaning and value to a designer's products, thereby ensuring that aesthetic innovations are translated into commercially viable products.

Presenting the city as a set of relations that are implicated in the development of fashion, however, does not negate a role for the individual designer. Within the fashion production process, the designer is not a passive recipient of influences from her institutional setting, but also an active shaper of that setting, in some cases challenging and altering its parameters. Numerous examples were raised in the course of interviews of how designers would assert their own individuality in response to constraints that were deemed restrictive. With respect to price constraints, for instance, one high-end designer sought a compromise between what was needed and what she desired in developing her lines. For her, "the line is normally a balance of different kinds of items, pieces you know are going to make money and then pieces that are the fringe and fluff, that give the line attitude" (personal interview, 2000). Rather than dismissing the influence of the designer within the fashion production process, the objective of this chapter, then, is to properly situate this agency within its broader institutional context. A narrative of the process by which New York fashion is produced illustrates how the city can *endow* the designer with competencies and resources for managing the often contradictory business of "commercial art." It suggests that a *recursive* relationship between the city and the designer (and, by extension, the design) – rather than a linear one – is the engine that drives fashion.

The narrative above also draws attention to the challenges involved in preserving this recursive relationship. The city's ability to function as a source of aesthetic inspiration has been increasingly eclipsed by the corporatization of key cultural intermediaries.[9] Trends such as the consolidation of retail buying power and acquisitions within the fashion magazine industry create a hierarchical production system in which commercially-oriented institutions have more leverage over design and marketing activities (Kucynski, 1999; WWD, 1999). In the case of the magazines for instance, the cost of one ad in *Vogue* is now upwards of $6,000, implying that entry is restricted to those designers who can exploit their economic – relative to creative – resources. In some cases, designers can bypass such intermediaries and disassociate themselves from a "corporate" identity by opening their own retail shops or by using public ads (see, for example, White, 1997). These options need to be encouraged and made more viable. As McRobbie (2002) suggests, cultural policy with respect to design should be centered on providing alternative means of ascribing value to designers, which are not solely – or even primarily – based on entrepreneurial or business criteria.[10] Developing and supporting a diversity of institutions – those driven by economic *and* cultural logics – is critical for safeguarding the role of the city as a source of cultural capital, and for maintaining the delicate balance between "art" and "commerce" that drives the fashion system.

Notes

1 For insightful studies on the relation between place and creativity, see Zukin (1995), Leslie (1997), Crewe and Beaverstock (1998), Grabher (2001), Molotch (2002) and Drake (2003).

2 This analysis is based on in-depth interviews conducted in November–December 1999 and January 2000 with over sixty industry actors, including high-end designers, retail buyers, and representatives from buying offices, forecasting services, buying offices, trade associations, design schools and other fashion-related services in New York City. I also draw on a survey of 28 high-end designers, conducted from December 1999 to March 2000.

3 The high-end segment consists of "couture," "designer," "bridge," "contemporary" and "better" categories, and it can be distinguished from the moderate segment in terms of design intensity, price and distribution channels, as high-end women's wear is more likely to be sold in specialty stores while moderate apparel tends to be sold in department stores (Uzzi 1993; Rantisi 2002b).

4 For an excellent review of the inherent tension between the artistic process and the demands of a commercial industry, see the study by McRobbie (1998) on the British fashion industry. An objective here is to illustrate how the city can negotiate this tension.

5 "7th" refers to Seventh Avenue, the main corridor in the Garment District, where the showrooms for the high-end designers are located. "6th" is the avenue where Bryant Park is located.

6 The analysis in this section draws on Rantisi (2002b).

7 More recently, resident buying offices, which link buyers to local designers, have also moved into the business of providing such services. They benefit from having an

intermediary position that exposes them to shifts in supply and demand. According to the forecasting director for the largest buying offices, "designers come to us because we spread the [research] costs among a large client base. We have an advantage as well by being part of a large buying office. We can track what sells and what doesn't" (personal interview 1999).

8 Ten apparel retailers control almost half of total retail sales, and among department stores, the top six chains control 90 percent of the sales (*WWD*, 1999).

9 McRobbie (2002) provides an account of how the climate of neo-liberalization has emasculated creativity in the U.K. fashion sector.

10 Santagata (2002) outlines some possible directions for such a policy orientation.

References

Agins, T. (1999a) *The End of Fashion*, New York: William and Morrow Company.

—— (1999b) "The Real Action in Fashion is Off the Runway," *Wall Street Journal*, 30 July 1999, B1, B4.

Bjorkegren, D. (1996) *The Culture Business*, London: Routledge.

Chase, E. W. and Chase, I. (1954) *Always in Vogue*, London: Gollancz.

Chen, J. (1999) "An Early U.S. Fashion Week," *CNN Style*, 14 September 1999; available online: <http://www.edition.cnn.com/STYLE/9909/14/focus.money> (accessed 16 June 2003).

Crewe, L. (1996) "Material Culture: Embedded Firms, Organizational Networks and the Local Economic Development of a Fashion Quarter," *Regional Studies*, 30: 257–72.

Crewe, L. and Beaverstock, J. (1998) "Fashioning the City: Cultures of Consumption in Contemporary Urban Spaces", *Geoforum*, 29: 287–308.

Drake, G. (2003) "'This Place Gives Me Space'; Place and Creativity in the Creative Industries," *Geoforum*, 34: 511–24.

Fashion Center BID (2001) Available online: <http://www.fashioncenter.com> (accessed 1 December 2001).

Florida, R. (2002) *The Rise of the Creative Class*, New York: Basic Books.

Gardner, H. E. (1993) *Creating Minds*, New York: Basic Books.

Ghiselin, B. (ed.) (1996) *The Creative Process: Reflections on the Invention of Art*, Berkeley, CA: University of California Press.

Grabher, G. (2001) "Ecologies of Creativity: The Village, the Group, and the Heterarchic Organization of the British Advertising Agency," *Environment and Planning A.*, 33: 351–74.

Guilbaut, S. (1983) *How New York Stole the Idea of Modern Art: Abstract Expressionism, Freedom, and the Cold War*, Chicago: Chicago University Press.

Harlan, H. (2000) "Downtown Funky in New York City," *AsianWeek*, 28 September 2000; available online: <http://www.asianweek.com/2000_09_28/feature.html> (accessed 25 May 2003).

Harvey, D. (1989) *The Condition of Post-Modernity*, Cambridge, MA: Blackwell.

Helfgott, R. (1959) "Women's and Children's Apparel," in M. Hall (ed.) *Made in New York*, Cambridge, MA: Harvard University Press.

Hirsch, P. (1972) "Processing Fads and Fashions: An Organization-set Analysis of Cultural Industry Systems," *American Journal of Sociology*, 77: 639–59.

—— (2000) "Cultural Industries Revisited," *Organization Science*, 11: 356–61.

Holusha, J. (2002) "Gritty Neighborhood is Looking Better to Retailers," *New York Times*, 13 October 2002, Section 11: 6.

Howkins, J. (2001) *The Creative Economy: How People Make Money from Ideas*, London: Allen Lane.

Jackson, K. (1995) *The Encyclopedia of New York City*, New Haven, CT: Yale University Press.

Koestler, A. (1990) *The Act of Creativity*, New York: Penguin Books.

Kucynski, A. (1999) "Conde's Latest Acquisition Has Fashion Industry Fidgeting," *New York Times*, 23 August 1999, Section A: 1.

Lash, S. and Urry, J. (1994) *Economies of Signs and Space*, London: Sage.

Leadbeater, C. (1999) *Living on Thin Air: The New Economy*, London: Penguin Books.

Leslie, D. (1997) "Abandoning Madison Avenue: The Relocation of Advertising Services in New York City," *Urban Geography*, 18: 568–90.

Lubow, A. (1999) "The Shadow Designer," *New York Times Magazine*, 14 November 1999, 124–27: 140.

Malossi, G. (ed.) (1998) *The Style Engine*, New York: The Monacelli Press.

McRobbie, A. (1998) *British Fashion Design: Rag Trade or Image Industry?*, London: Routledge.

—— (2002) "Fashion Culture: Creative Work, Female Individualization," *Feminist Review*, 71: 52–62.

Molotch, H. (2002) "Place in Product," *International Journal of Urban and Regional Research*, 26: 665–88.

Museum of the City of New York (2000) "Fashion on Stage: Couture for the Broadway Theatre, 1910–1955"; available online: <http://www.mcny.org/Exhibitions/prevexhib.htm> (accessed 20 May 2003).

New York Industrial Retention Network (2001) *Still in Fashion: The Midtown Garment Center*, special report by the New York Industrial Retention Network to the Union of Needletrades and Industrial Textile Employees (UNITE), April 2001, New York.

Owen, B. (1987) *Costume Design on Broadway: Designers and Their Credits 1915–1985*, Westport, CT: Greenwood.

Pratt, S. (1997) "Head to Head: Elizabeth vs. Orchard Street," *New York Post*, Sunday edition, 3 August: 34.

Rantisi, N. M. (2002a) "The Local Innovation System as a Source of Variety: Openness and Adaptability in New York City's Garment District," *Regional Studies*, 36 (6), 587–602.

—— (2002b) "The Competitive Foundations of Localized Learning and Innovation: The Case of Women's Garment Production in New York City," *Economic Geography* 78: 441–62.

—— (2004) "The Ascendance of New York Fashion," *International Journal of Urban and Regional Research*, forthcoming.

Santagata, W. (2002) "Cultural districts, Property Rights and Sustainable Economic Growth," *International Journal of Urban and Regional Research*, 26: 9–23.

Scott, A. J. (1996) "The Craft, Fashion, and Cultural Products Industries of Los Angeles: Competitive Dynamics and Policy Dilemmas in a Multi-Sectoral Image-producing Complex," *Annals of the Assocation of American Geographers*, 86: 306–23.

—— (1997) "The Cultural Economy of Cities," *International Journal of Urban & Regional Research*, 21: 323–39.

—— (2000) *The Cultural Economy of Cities*, London: Sage.

—— (2001) "Capitalism, Cities, and the Production of Symbolic Forms," *Transactions of the Institute of British Geographers*, 26: 11–23.

—— (2002) "Competitive Dynamics of Southern California's Clothing Industry: The

Widening Global Connection and its Local Ramifications," *Urban Studies*, 39: 1287–306.

Scott, W. B. and P. M. Rutkoff (2001) *New York Modern: The Arts and the City*, Baltimore, MD: Johns Hopkins University Press.

Scranton, P. (1998) "From Chaotic Novelty to Style Promotion: The United States Fashion Industry, 1890s-1970s," in G. Malossi (ed.) *The Style Engine*, New York: The Monacelli Press.

U.S. Bureau of the Census (1997) *Economic Census 1997*, Washington, DC: Government Printing Office.

U.S. Bureau of Labor Statistics (2001) *Current Employment Statistics, 2001*; available online: <http://www.stats.bls.gov> (accessed 5 December 2001).

Uzzi, B. (1993) "The Dynamics of Organizational Networks: Structural Embeddedness and Economic Behavior." PhD dissertation, New York: State University of New York at Stony Brook.

White, C. C. R. (1997) "New Wave of Designers Opening Stores in Soho," *New York Times*, 2 September 1997, Section C: 19.

WWD (1999) "NPD: Department Stores Taking a Bigger Slice of Apparel Pie," *WWD*, 30 April 1999: 4.

Zukin, S. (1991) *Landscapes of Power: From Detroit to Disney World*, Berkeley, CA: University of California Press.

Zukin, S. (1995) *The Culture of Cities*, London: Blackwell.

7 Creative resources of the Japanese video game industry

Yuko Aoyama and Hiro Izushi

Introduction

Although the video game industry may not generally be regarded as a sector offering forms of "haute culture" or refined art, it has become a sizeable industry with pervasive influences on popular culture. The worldwide market for video games is estimated to be at $18 billion. In 1999 alone, 215 million copies of games were purchased in the U.S., which amounts to two for every American household (IDSA, 2001). Despite its ubiquity and its significant influence upon youth culture, however, relatively little is known about the industry and, particularly, the role of creative resources in industry formation and competitive strength. Research on video games so far has emphasized the moral issues and psychological impacts (Loftus and Loftus, 1983; Greenfield, 1984; Kinder, 1991; U.S. Congress, 1994). The industry's developmental trajectory, which incorporated multiple aspects of emerging technologies in computers, multimedia and the Internet, represents a contemporary synergy of digital technologies, artistic creativity and multimedia entertainment.

Broadly speaking, there is an increasing domination of U.S. and English language-based exportable cultural products, under the influence of multinational firms of the U.S. origin, in the form of Hollywood films and rap music, and in commodities such as Levi's jeans, Coca-Cola and Nike shoes, forcing other national players into a niche market (Llewelyn-Davies *et al.*, 1996). Japan's video games therefore represent an exception to this trend in the Western world. Although the industry was first established by Atari, an American firm, most industry insiders today agree that without Japanese firms such as Nintendo and Sony Computer Entertainment (SCE), it would not have amounted to more than just a passing fad. In fact, after the famous video game industry crash of 1983, which virtually wiped all demand in the U.S., Nintendo played a significant role in re-establishing the industry as a profitable business. More recently, Sony, armed with significant reputation in consumer electronics, successfully combined hardware and software resources to dominate the global market with PlayStation and PlayStation 2. How did the Japanese video game industry manage to penetrate the global market?

This chapter explores the above question by addressing widespread misconceptions that surround Japan's video game industry. While we acknowledge numerous books that document industry trends available in the Japanese language, few scholarly studies exist that focus on the cross-sectoral use of creative resources and offer a systematic interpretation on the role of technical expertise developed in other industrial sectors.[1] Our aim here is to analyze this relatively neglected aspect of the industry through examining its distinctive historical, institutional and cultural foundations.

We argue creative foundations in cartoons and animated films, as well as cross-industry links to consumer electronics, have functioned as important foundations for the emergence of this industry in Japan. We begin by discussing the significance of and misconceptions about video games as an industry. Next we analyze how its evolutionary trajectory was influenced by interplays between hardware manufacturers and software publishers. We then shift our discussion to the skills foundations of the industry, based on historical and contemporary evidence for inter-industry sharing of labor pools and drawing from interviews of industry insiders.

Video games as a creative industry: its significance and misconceptions

As Walter Benjamin (1936) argued during the industrial revolution, new technologies in any era deeply influence art form of the period. Advanced information and communication technologies now offer a new frontier through digitization and simulation, opening up possibilities for new forms of art in cyberspace. Unlike cinema and television, which offer passive entertainment, video games emerged out of older interactive entertainment such as toys and arcade games, and expanded interactivity to a new realm by allowing players to adopt a character, function as a protagonist, craft story-lines, engage in role-playing and even interact with those from opposite ends of the world (Jenkins, 2000; Wolf, 2000).

As a contemporary art form, video games occupy an important position in the already fledgling creative industries in advanced economies. With the current world market of U.S.$18 billion, markets in G-5 countries are projected to continue growing up to 2005. Already in 1997, video games were more ubiquitous in the U.S. than home-based Internet access, and in Japan 78 per cent of all households own video games (CESA, 2000). Video games and related industries generated 220,000 jobs and U.S.$7.2 billion in wages in the U.S. in 2000 (IDSA, 2001). The video game hardware and software alone generates 30,000 jobs in Japan and 27, 000 jobs in the U.K. (METI, 2001a; Department for Culture, Media and Sport, 1998).

Despite its ubiquity and its significant influence upon youth culture, however, relatively little is known about the industry's evolutionary path, internal dynamics and the role of creative resources in innovation. Particularly little is understood of the reasons behind the dominance of Japanese firms. In addition to Nintendo and SCE, there is a plethora of Japanese software publishers. To better understand the competitive position of Japanese firms, we need to first address some widespread misconceptions surrounding Japan's video game industry.

First, there is a general consensus that Japan's competitive strengths in international trade lie in manufacturing industries (e.g. automotive, electronics, machinery) but not in creative, knowledge-based industries. The country's weakness in the latter is evident particularly in its uncompetitive software industry. It has been suggested that Japan's institutional arrangements in trade, finance, employment and education, with their emphasis upon conformity, loyalty and stability, suppress creativity and risk-taking entrepreneurship and thus constitute the source of its weakness in software (Cottrell, 1994; Anchordoguy, 2000). Given the institutional arrangements hindering the development of creative resources necessary for software design, how have Japanese video game firms succeeded in developing their products? Archordoguy (2000) argues that game software, which constituted 27 percent of packaged software sales in Japan in 2000, is distinctive from other software and therefore should be dealt with separately. While software is generally language-driven, hardware manufacturer-oriented, and targeted for personal and business communication, she argues that game software is part of a new industry with lower barriers to entry, does not suffer from language-based disadvantages and escaped the negative aspects of state-led industrialization. Unlike Japan's other successful industries, the video game industry indeed did not receive governmental support, and it evolved solely out of private-sector initiatives. Yet, we find these explanations insufficient in understanding the rise of Japan's video game industry. For one, the argument for entry barriers may not work the same way for creative industries. A study by Bryman (1997), for example, showed that late entrants had significant advantages over industry pioneers in case of the U.S. animation industry. In addition, as we shall show, significant barriers to entry were observed for platform developers as well as for software publishers. Most importantly, however, the above explanations fail to identify creative resources for video games. What creative resources, if any, have Japanese video game firms made use of?

Second, creative industries have traditionally been regarded as craft-based, low-tech and labor-intensive: fashion, drama, arts. More recently, however, links to technology-intensive industries, particularly with the introduction of multimedia technologies, have become an integral aspect of creative industries, and video games are the prime example. Evidence from the U.S. indicates that the video game industry typically shows a higher R&D to sales ratio among entertainment industry (IDSA, 2000). One programmer we interviewed even argued that the cutting-edge in graphical interfaces, etc. is in the video-game industry, surpassing other sectors such as computer-aided manufacturing and design.

Video games represent interactions between hardware manufacturing and software publishing, the significance of which remains ambiguous. It has been suggested that the co-location, or the lack of co-location, of creative/software industries and hardware industries matter to the development of the former

(Pratt, 1997). Quite separately, von Hippel (1988) argued that users' knowledge about their tasks and needs can give rise to the formation of specialized complexes of software firms. This raises the question of whether the development of this high-tech creative industry in Japan linked to the country's resources in hardware technology.

Recent theoretical developments offer various insights into forms of industrial organization, with separate approaches such as that of transaction costs (Coase, 1937; Williamson, 1975), socio-institutional foundations (Piore and Sabel, 1984) and embeddedness (Granovetter, 1985). Nelson and Winter (1982) advocated an evolutionary approach, in which "path dependency" is observed through technologies' developmental trajectories. More recently, Osborn and Hagerdoorn (1997) suggested a pluralistic approach that allows a better understanding of the complexity in organizational behavior. In our view, a pluralistic approach best accommodates distinctive characteristics of cultural industries. Our historical analysis of video games emphasizes the evolutionary process through which cross-sectoral transfers take place concurrently with the evolution of technological knowledge and artistic creativity.

We argue that the industry is embedded in a social and historical foundation of creative imaginary based on vibrant cartoons (*manga*) and animated films, complemented by the availability of skilled engineers from Japan's consumer electronics industry. The presence of a competitive consumer electronics industry in Japan meant that a foundation of necessary skills and labour existed; and social legitimacy drawn from pre-existing comic and animation industry has served as a foundation for game creators and designers. These foundations, combined by the timely entry of platform developers such as Nintendo and Sony, played an important catalytic role in the emergence of the video games industry, and resulted in a vertically disintegrated industrial complex with hardware manufacturers as well as a plethora of small game-software houses.

Cross-sectoral links and sharing of resources for video game development

The evolution of the industry has been influenced by a variety of actors from broad sectoral backgrounds. The onset of the industry, which began with the application of computer programming for arcade games, was subsequently infused with technologies in consumer electronics to produce home-based video game consoles. The battle over platforms has been increasingly based on access to hit game software. In Japan, the two major platform developers, Nintendo and SCE, forged an indirect link between manufacturing expertise in consumer electronics and software publishers. As hybrids of hardware manufactures and software developers, platform developers drew crucial resources indispensable for growth from consumer electronics, and helped fledgling software start-ups with financial assistance and early disclosure of platform specifications.

Links between consumer electronics manufacturers and platform developers

As mentioned earlier, technical capability and expertise in consumer electronics have served as a critical foundation for the early development of platforms in Japan. Nintendo's initial shift into the video-game business was made possible through collaborations with consumer electronics firms, and the sharing of engineering skills with them. The local presence of consumer electronics firms, and particularly IC chip manufacturers, such as Mitsubishi Electric, Ricoh and Sharp, enabled frequent communications and experiments during the initial stages of new console design. Nintendo's interest in video games in the mid-1970s coincided with that of Mitsubishi Electric, which was in search of a new partner after a joint LSI (large-scale integrated circuits) development for video games collapsed as a result of its partner's bankruptcy. The collaboration resulted in Nintendo's successful launch of the first color home-based video game console ("Color TV Game 6"), and provided an opportunity for Nintendo engineers to receive hands-on training in LSI design. Nintendo's subsequent alliance with Ricoh in their development of Famicon was in part motivated by an ex-Mitsubishi Electric engineer at Ricoh who had worked with Nintendo on LSI design for video games. Sharp had been a partner in previous electronic toy ventures, and the relationship nurtured during the previous project resulted in Sharp providing the liquid crystal display technology that powered Nintendo's early success with hand-held video games ("Game & Watch"). Although Nintendo now deals with overseas manufacturers including Silicon Graphics, Philips and IBM, collaborations with domestic electronic firms were critical to Nintendo's early models.

In addition to supplying technological knowledge through sharing engineers, consumer electronics firms played a critical role in Nintendo's success by serving as contract manufacturers. Apart from its sales offices around the world, Nintendo, headquartered in Kyoto, maintains no in-house plant, and only has a warehouse with product-checking functions near Kyoto. Nintendo outsources most manufacturing activities, including the production of parts and the assembly of platforms and cartridges. In the early days when sophisticated international supply chains were still underdeveloped, the use of domestic consumer electronics firms as contractors helped Nintendo concentrate its resources on the design and development of platforms.

SCE's successful entry into the market with CD-ROM technologies represents a different relationship between the consumer electronics industry and platform developers. As a reputable consumer electronics firm itself, SCE uses Sony's in-house plants known for quality in design and performance. Direct control over the production of consoles, coupled with faster production, allowed SCE to remain responsive to fluctuating market demands while maintaining quality standards. Yet another model is that of Sega; although now defunct, Sega exploited close links to resources possessed by an agglomeration of small engineering firms in Tokyo's Ohta-Ward that is known for swift and flexible production of parts and components for prototypes.

Japan's video game industry has therefore benefited from technical knowledge and engineering skills found in its globally competitive consumer electronics industry. In addition, a variety of other industries provided skills resources, such as programming from automobiles, computers and telecommunications, multimedia technology from broadcasting, and system design from electronics industry, as well as CAD engineering from the construction industry (Sunagawa, 1998).

Links and cultural proximities between platform developers and software publishers

Both Nintendo and SCE today have extensive in-house resources for software publishing as well. The co-existence of hardware and software businesses creates a degree of synergy that is deemed critical in the success of the two firms. For instance, at Nintendo's software development division, all staff members, including designers and artists as well as software programmers, are given training in hardware architecture. Shigeru Miyamoto, director of the division, argues that a combination of knowledge about hardware and software allows systematic development of games at all levels within his teams (Takeda, 1999: 100–101). Yet, the Japanese video-game industry consists of a plethora of independent software publishers as well as the two platform developers, forming a web of links that transfer expertise. As shown in Table 7.1, software publishers vary in origin, size and corporate organization. There are broadly categorized into three groups: 1) in-house publishers of platform developers (e.g., Nintendo, SCE); 2) comprehensive software publishers with in-house capability in most aspects of software development from scenario writing to score writing and sound design (e.g. Square); and 3) publishers that act as producer/coordinator and outsource most functions (e.g. Enix).

Albeit indirect, links exist between third-party software publishers and the consumer electronics, in which platform developers function as a medium. As hybrids of hardware and software businesses, platform developers facilitate the entry and growth of independent software publishers by taking the lead in information exchange in the following three stages: 1) initial guidance on software formats, quality and ethics standards, purchase of development tools, etc.; 2) provision of new versions of operating systems during the development of game titles at software publishers; and 3) final quality inspection on software and documentation.

Evidence shows a presence of cultural proximity between platform developers and third-party software publishers. Japan's software publishers have been in a unique position to access, almost exclusively, dominant platform developers particularly at the early stages of the industry's development. There are multiple ways in which cultural proximity matters, yet in all cases it functions to reduce communication barrier and facilitates flows of information. Firstly, the presence of face-to-face personal networks allows for informal contact that ensures speedy access to and disclosure of formerly proprietary information. Platform developers normally disclose new platform specifications to third-party software

Table 7.1 Major software publishers in Japan

Company	*Year established*	*Origins*	*HQ location*	*Employees*	*Sales* (FY2000) in millions U.S.$*	*Major titles*
Bandai	1950	Toys	Tokyo (Taito)	960	1,808	Dragon Ball
Capcom	1979	Arcade games	Osaka	1,009	409	Resident Evil, Biohazard, Onimusha
Enix	1975	Publishing	Tokyo (Shibuya)	130	377	Dragon Quest
Koei	1978	PC sales, office software	Ashikaga → Yokohama	435	167	Dynasty Warriers, Winbank
Konami	1973	Jukebox rental/repair	Tokyo (Minato)	522	1,429	Metal Gear Solid, Silent Hill
Namco	1955	Arcade games	Tokyo (Ohta)	2,097	1,221	Pac-Man, Tekken
Nintendo	1889	Toys	Kyoto	1,150	3,854	Mario Brothers, Pokémon
SCE	1993	Consumer electronics/Recording	Tokyo (Minato)	1,700	N/A	Gran Turismo, Spyro, Cool Boarders
Sega	1951	Jukebox mfg	Tokyo (Ohta)	1,007	2,024	Virtual Racing, Virtua Fighter
Square	1986	Game software	Tokyo (Meguro)	774	608	Final Fantasy

Source: Compiled by authors from company websites

Note
*Total corporation-wide sales. Converted from yen at 120 yen per U.S. dollar

publishers at middle to late stages of development. By doing so, platform publishers can incorporate feedback from software publishers before the final completion. It also helps software publishers to resolve any compatibility problems. Through such interactions, platform developers aim at the timely delivery of new titles as well as the successful adaptation of existing titles to the new platform. Occasionally, platform developers distribute a small number of prototype consoles to a few selected software publishers with close business relationships two to four years in advance of the planned release. Such practices give Japanese software publishers a head start in advance of the release of new platforms.

The second aspect of cultural proximity refers to the presence of common operational procedures. For example, comparing software publishers in the U.K. and U.S. as a partner for joint development, Shigeru Miyamoto, the father of Nintendo's "Donkey Kong" and "Mario Brothers," finds it far easier to work with U.K. software publishers due to the similarity of their approaches to product development. In Miyamoto's view, U.K. software publishers pay attention to the structure of games at all stages of the development process so that the games can be programmable. Their approach is based on a solid understanding of hardware architecture, widely shared by Japanese software publishers, but seldom found among U.S. publishers.

Finally, the third form of cultural proximity may be observed between software publishers and the market in terms of demand-responsiveness. As a content-driven commodity, local game publishers enjoy the advantage of knowing the preferences and tastes of the domestic market in detail. Popular genres of games vary significantly between countries (Monopolies and Mergers Commission, 1995: 48); while two-thirds of games sold in the U.S. in 1999 were either action or sports games, only one-third of the games sold in Japan that year fell into the same categories (Shintaku and Ikuine, 2001). The "Army Men" series developed by American publisher, 3DO, successfully cater to boys who also play with plastic toy figures, but overseas markets presumably will not respond in the same manner. In Japan, role-playing games enjoy huge popularity, some being specifically targeted for the Japanese market. As it will be shown in the following section, interaction between video games and cartoons/animated films has created in Japanese game software a unique style that emphasizes characters and stories.

Sources of creativity in Japan's software publishing: cartoons and animated films

Today, software publishers command a decisive power in Japan's video game market. While the history of the industry is dominated by competing platforms, an increasing proportion of profits is being made in game software. Platform developers sell their consoles at a razor-thin margin and make a greater part of their profits from in-house software publishing or licensing fees from third-party software publishers. The ratio between hardware and software markets in Japan shifted from 1:2.4 in 1990 to 1:3.2 in 1998. Further, platforms are being made obsolete as software distribution will increasingly take place over the Internet in the future. This trend, while still limited at this time, will further push software publishers to the center stage of the industry.

Production and the division of labor in software publishing

The production of game software requires a variety of technical skills and artistic creativity. Individuals work as part of a multi-disciplinary team and are required to interact with others from a variety of backgrounds. Table 7.2 shows some of the major occupational categories in the industry, with skills varying from design and music to modeling and programming. Since its infancy in the U.S., the video-game industry has offered employment opportunities for artists and programmers who prefer to work in an "alternative-to-the-mainstream" environment. It attracted engineers and programmers, many of whom were nurtured in close association with the country's defense interests under the Cold War regime. Bushnell, the founder of Atari, was quoted as saying, "we provided a place for creative people to be part of something completely new. These were people who wanted to create something intellectually stimulating and fun. They wanted to put their talent into making games, not bombs" (Sheff and Eddy, 1999: 140). Japan's software publishers today also showcase their work environment as being casual and informal, creativity-oriented and divorced from the strict hierarchy of firms in traditional industries.

Today, the production of game software consists of a number of stages (see Figure 7.1). The production starts with planning of game titles, including

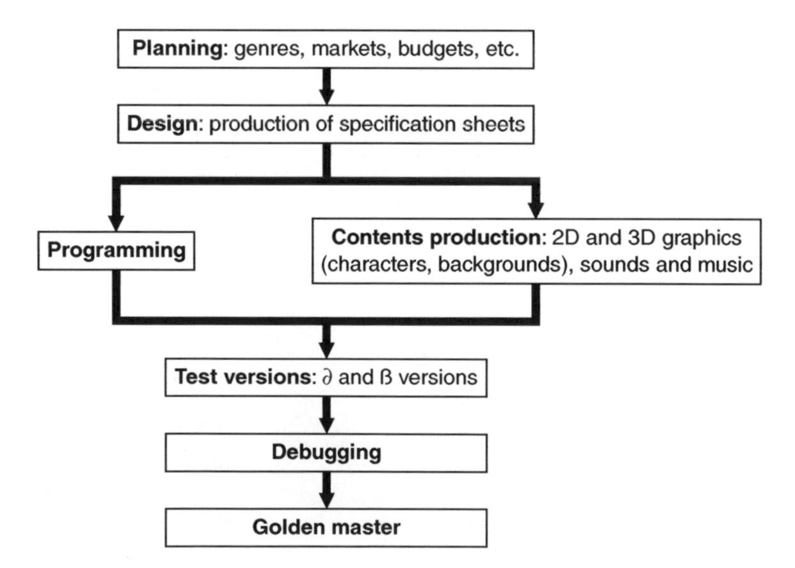

Figure 7.1 Production process of game software (Source: Ueda (1998), Dai-X Shuppan (2000), Yanagawa and Kuwayama (2000))

Table 7.2 Occupational categories in the video-game industry

Occupational category	*Summary of major tasks*	*Skills/training required*
Game Designers	Write the blueprint of the game; decide the mission, theme and rules of play; lead game designer determines overall concept, level designers work on smaller sections of the game, and writer-designers write text and dialogue.	Writing and communication skills; management skills; technical skills including computer programming and software design; most designers earn a college degree in English, art, or computer science.
Artists	Create graphics; in a 2-D game, artists draw images on paper and then scan them into the computer, in a 3-D game, artists create most of their images within the computer; character artists design and build creatures, animators make them move; background artists create video-game settings; texture artists add detail to the surfaces of 3-D art.	Visual imagination; ability to apply basic math concepts; ability to use modelling and animating software; most video-game artists have formal training in fine arts or art-related subjects, bachelor's degrees are an advantage.
Sound Designers	Compose music and sound in a game; research appropriate music options for particular games; choose creative sound effects with the correct balance of realism and entertainment value of exaggeration.	Musical creativity; training in audio engineering; knowledge of the basics of computer hardware and software; many sound designers have a bachelor's degree in music and some have education in film scoring.
Programmers	Plan and write video-game software; translate ideas into mathematical equations the computer understands; specialists include engine programmers who write software that makes video games run, artificial intelligence programmers who write code to make computer-controlled characters act realistically, graphics programmers, sound programmers, and tool programmers who write software for artists and designers.	Strong math skills; knowledge of computer programming (C and C++); most programmers have a bachelor's in computer science.
Game Testers	Play games to find errors in software, graphic glitches, computer crashes and other bugs; write reports describing the problems they find.	Ability to communicate clearly; familiarity with technology and expert game-playing skills; does not require formal education but testers are encouraged to earn computer technician certificates.

Source: Adapted from Crosby (2000).

missions, themes, key characters, target platforms, market and the budget. Game blueprints are either made in-house or brought to publishers by freelance game designers. Upon being approved by the producer, who is responsible for the overall management of the project, production moves onto the specifics that include scenario writing, designing of character details and accessory items, and developing plans for backgrounds settings. This stage is followed by two concurrent tasks: production of graphic materials including characters and backgrounds, and programming. For graphic material production, characters are normally hand-drawn first by artists and then scanned into the computer with 2D or 3D software specially designed for graphics and animation. Sound and music are generated by sound designers. Computer programmers work on programming to define movements and backgrounds. Programmers then import graphic and sound materials and complete what is known as an "alpha version." At the next stage, debugging, game testers play the games and identify bugs, glitches and errors. At some point of the debugging stage, a beta version is often issued and made available to critics and mass media for pre-release evaluation and marketing. The production is complete when a "golden master" is issued and sent to mass production.

Over the course of its history, skills requirements in software publishing have witnessed two fundamental changes. The first took place when Nintendo introduced characters and story-lines, a significant development from the previously dominant "war game" and "ball game" types in which speedy reflexes determined the winner. Early video games, such as Russell's "Space War" (1961), Bushnell's "Pong" (1973) and Taito's "Space Invaders" (1978), all fall into this category. Nintendo's hit titles such as "Donkey Kong" and "Mario Brothers" introduced identifiable human/animal characters with scenarios and story-lines. These developments made skills for creating absorbing stories and appealing characters a necessity in video-game software publishing. Consequently, game design, which was largely within the realm of engineers until then, henceforth became the task of specialized designers with artistic talent. Engineers with high levels of mathematics increasingly specialize in middleware, e.g. proprietary programming software that transfers graphic data from industry-standard 2D/3D tools, such as Avid Technology's Soft Image, and optimizes them to a specific platform.

The second change took place with the shift from ROM cartridges to CD-ROM as the primary storage media. This development accelerated the already on-going shift in emphasis toward talented artists with superior graphics capability. Also, CD-ROM's large memory capacity allowed some game software to begin including pre-rendered movies to be shown during the game, increasing the need for more realistic movement by associated game characters. These two trends increased the importance of graphic-design skills further. To respond to these changing skills requirements, Japan's video-game industry drew artistic creativity from its already well-developed cartoons and animated film industry.

Manga and animated films industry in Japan

Cartoons and animated films have a pervasive influence upon Japan's culture and society, far greater in comparison with American and European counterparts. Japanese cartoons, known as *manga*, comprised over one-third of all books/magazines sold in 1999 (Dentsu Soken, 2001). There were 2,192 titles of new *manga* books published that year, in addition to 280 titles of *manga* magazines with an annual circulation of 1.1 billion copies. Approximately two-thirds of boys and more than one-sixth of girls aged between 5 and 18 read *manga*, and most read weekly, biweekly or monthly comic magazines averaging 400 pages with 15 serialized stories (Schrodt, 1983; Grigsby, 1998). *Manga* are for adults, too; out of 280 comic magazines in circulation in 1999, 215 were magazines targeting adults with an annual total of 727 million copies printed comprising 53 percent of all comic magazines (Dentsu Soken, 2001). The subjects range from romantic to educational, humor, sports, adventure, sex and violence. There are even how-to *manga* (e.g. cooking, finance), *manga*-fied classics and socio-political satires.

Manga has a distinctive style, which owes much to the famous cartoonist known as the "god of *manga*," the late Osamu Tezuka (1926–89) (Bendazzi, 1994; *The Economist*, 1995). Influenced by Hollywood films at a young age, Tezuka revolutionized *manga* by incorporating techniques used in films, such as close-ups, fade-outs and varying camera angles. He has been influential in introducing new formats as well as contents, making *manga* develop from previously dominant short-strip humor to long and complex stories. He also succeeded in weaving in serious issues, such as religion, race, war and social justice, without leaving the boundaries of entertainment. Throughout his over 40-year career, Tezuka drew over 150,000 pages, with a record 250 million copies of sales by the mid-1980s. Tezuka was also the pioneer in producing an animation TV series in 1963, with his hit *manga* "Astro Boy." When Tezuka died in 1989, Japan's influential daily, *Asahi Shimbun*, carried an editorial that stated among many reasons for the popularity of *manga* in the country, the most important factor was that "Japan had Osamu Tezuka." Almost all professional cartoonists today are under his influence in one form or another. With the quality of his works as well as his credentials as a medical doctor, Tezuka was most influential in raising the status of *manga* in Japan to the level of literary works.

The importance of *manga* in Japan's popular culture meant high social status for cartoonists, who are regarded as celebrities with recognition that at least equals popular novelists.[2] Japan's comic artists typically own the copyrights to their series, and therefore enjoy greater control over their products than American serial-comic-book artists (Reinmoeller, 1999). Regarded as artist-entrepreneurs, the high social status associated with Japan's cartoonists is in part a reflection of their economic status. In 1997, there were 3,500–4,000 publishing cartoonists in Japan, in addition to 20,000 assistants (Dentsu Soken, 1999). Almost every educational institution from elementary schools to universities has student-run cartoonist clubs as part of extra-curricular activities. Ask Japan's third graders about their future career aspirations, and you will find a significant number aspire to become cartoonists. As a result, more than 10,000 individuals regularly attempt to become professional cartoonists in Japan (Lent, 1995). Comic books gave rise to animated films and TV series, further reinforcing the strength of *manga* as an integral part of popular culture in Japan. Most of the early TV animation programs in the 1960s and 1970s were based on characters and stories out of comic books. In the late 1970s and the 1980s, animated films emerged as blockbusters in the Japanese cinema, with such hits as *Cosmic Warship Yamato*, featuring a spaceship transformed from a World War II battleship, *Galaxy Express 999*, an adventure story of a boy traveling on a space train, and *Mobile Suit Gundam*, hard science fiction chronicling a war between people living in the earth and space colonies. Animated films and TV programs today account for a significant portion of Japan's market for films and broadcasting. For instance, of the revenues grossed by the top 10 films released in 1997, 40 percent were accounted for by animated films.[3]

Skills transfer from *manga* to animated films was clearly one of the consequences. Reiji Matsumoto, who produced such animated films as *Yamato* and *999*, was trained as a cartoonist with a strong influence from Tezuka. Another producer of the science-fiction animation series *Gundam*, Yoshiyuki Tomino started his career at Tezuka's firm, and Hayao Miyazaki, who produced *Spirited Away*, a 2003 Academy Award winner, also produces *manga* and does not hesitate to acknowledge Tezuka's strong influence in his work.

Transfer of creative resources to video games

Cartoons and animated films provided an important foundation for the emergence of Japan's video-game industry. The core of game software production involves scenario writing and drawing, and cartoons and animated films provided the necessary skills and expertise for character production and graphic design. The higher incomes offered by the video-game industry encouraged a migration of skilled designers and illustrators from cartoons and animation. In the animation industry, the hierarchical structure with a series of small subcontractors tends to suppress wages, and project-based fees often reduce hourly wages.[4] The video-game industry, which enjoyed rapid market growth in the 1990s, managed to employ workers at higher wages, and even caused a labor shortage in the animation industry (Nikkei BP, 1999).

While no data exist on the actual transfer of employment between cartoons, animation and video games, the presence of interactions between these sectors are clear from literature on the video-game industry, and was also supported by our interviews with industry insiders.[5] The overlap in required skills between cartoons/animated films and video games can also be observed through programs in vocational schools. There are 300 vocational programs that offer training specifically designed for video-game-related occupations, and an additional 170 vocational programs offer training for those who aspire to become cartoonists, animators and illustrators. These vocational schools are typically classified into two categories: fashion/art/design schools, and electronics/computer programming schools (see Table 7.3). Among the 12 vocational schools surveyed for this research in Tokyo, the majority fell into these two categories, with the exception of two that were specialized schools devoted to developing a variety of skills for video game production. In all arts and design schools surveyed, programs on animated films and cartoons are run in conjunction with ones on video games, often sharing curricula (e.g. computer graphics, 3-D animations) between them.

The video-game industry maintains close ties to these vocational schools. Practitioners in the industry often serve as instructors, and occasionally students are hired as part-time workers, an entry-way for full-time employment later. Some firms go as far as setting up their own vocational schools; platform developers as well as major software publishers jointly established Digital Entertainment Academy in Tokyo, and Konami runs its own, Konami Computer Entertainment School in Tokyo and Osaka. These schools are popular among aspiring game designers because they open doors to employment in a highly competitive job market in the video-game industry (Mikuriya, 1999).

The sharing of creative resources in Japan was in part possible due to the contemporaneous and overlapping development of cartoons, animated films and video-game industries. Unlike the case of the U.S., where the influence of comic books and animated films in popular culture peaked in the immediate post-war period and had already waned by the time video games emerged, Japan's comic-book publishing peaked as recently as the mid-1990s.[6] Animated films in Japan today occupy the major share of the otherwise seriously declining domestic film industry. Skills resources in these industries were concurrently developed and far more effectively shared in Japan, thereby increasing cross-influences of contents among these creative industries.

The sharing of artistic talent between cartoon/animated films and video games in Japan has led to a number of common features that can be observed between the two forms of art. For example, many in the video-game industry refer to their interest in cartoon/animated films as being the primary motivating factor for pursuing current occupation. In fact, Nintendo's innovation of incorporating human and animal characters and stories in its video games, as noted earlier, was attributed to an individual designer who was also under Tezuka's influence. Shigeru Miyamoto, Nintendo's chief game developer and the creator of "Donkey Kong" and "Mario Brothers," began his interest in *manga* in junior high school, and led the cartoonist club at school. He later attended Kanazawa College of Art and was hired by Nintendo as its first staff artist in 1977. When he was assigned to develop an arcade game for the U.S. market in 1980, he studied the then-popular "Space Invaders" and "Pac-Man." Upon realizing that the objective of these games was simply to erase objects from the screen, he decided to create a game with a story just as in cartoons/animated films.

In the case of "Donkey Kong," the simple game is built upon a detailed *manga*-like character with an elaborate story that resembles those from

Table 7.3 Examples of vocational schools in Tokyo offering courses on video games

Types		*Courses on video games*	*Program duration: Enrolment*	*Other courses*
Specialized video-game training schools	1	game programming, planning, graphics, sound design	2 years: 230 3 years: 50	none
	2	game programming, CG, sound design, planning, scenario writing	3 levels (1 level/year): total of 250	none
Art/design schools	3	game programming, game design, character design	2 years: 120	cartoons (2 years: admissions 80), animated films (2 years: 80), broadcasting, IT, architecture
	4	game design, CG design	2 years: 200	cartoons (2 years: 150), animated films (2 years: 250), voice acting, illustration
	5	game graphics (as part of CG design)	2 years: approx. 40	cartoons (2 years: approx. 40), graphic design, interior design
	6	game design, character design, game sound, game production	2 years: approx. 200	cartoons and animated films (2 years: approx. 200), graphic design, fashion design, interior design, architecture, industrial design
	7	game programming, game graphics, mobile-game programming	2 years: 120	cartoons (2 years: 50), animated films (2 years: 40), 3D movies, Web design, voice acting
Electronic/ Computer-programming schools	8	game design and programming	2 years: 40	information processing, IT, architecture
	9	game programming, game CG	2 years: 80	information processing, architecture, biotechnology, environmental technology
	10	game design and programming	2 years: 30	computer graphics, architecture, information
	11	game programming, game CG (as part of CG course)	3 years: 70	processing system engineering, computer programming, computer music
	12	game design and programming	2 years: 40	3D computer graphics, web design, digital movies, information processing, networks management

Source: Authors' survey

animated films, "in which the lead character overcomes a variety of barriers and wins back his princess from a mischievous gorilla" (interview with Miyamamo quoted in Aida and Ohgaki 1997: 178). Miyamoto spent time on character development, to such an extent that the lead character is made as specified as a blue-collar construction worker who keeps a gorilla as a pet and has a pretty girlfriend. The same lead character was employed again, appearing as "Mario" in "Super Mario Brothers."[7] Video games have since become far more complex and elaborate. As an extension of such early games as "Donkey Kong," role-playing games, in which players can pick and assume a character and typically go on an adventure or treasure hunt, enjoy broad popularity in Japan. Some of the role-playing games feature characters out of *manga* and animated films, or with features resembling them, suggesting that drawing techniques and character design are directly transferred from *manga* or animation film industries.

The three media, video games, cartoons and animated films are increasingly linked with one another in their marketing to boost sales of one another. For example, a few classic animated films have been adapted for video games. Examples include *Space Warship Yamato*, released for PlayStation in 1999, and *Mobile Suit Gundam* for PlayStation 2 in 2000. The import of classic animated films, such as *Yamato* and *Gundam*, is also found in typing training programs on the PC format. These games include video clips from the originals and use the same voice actors to capture those grown-ups who used to watch the originals. Yet the best-well-known example of the three media's integration is the success of Pokémon. Pokémon was first developed as a video game for Nintendo's Game Boy. Shortly after its debut in 1996, the character began appearing in monthly *manga* magazines for children. This ignited its popularity, resulting in a TV animation program series in 1997, followed by a couple of blockbuster films. The appearance of Pokémon in a variety of media reinforced one another, facilitating the growth of the market for video games.

Conclusion

The developmental trajectory of the video-games industry revealed a complex interplay between hardware and software technologies from its origin, and more recent trends represent a transition from a hardware, engineering-driven to an increasingly software-centered industry supported by artistic creativity drawn from cartoon and animation film industries. This shift can be clearly seen in the fact that the battle over platforms today to a large extent hinges upon the coordination with software publishers that provide hit game titles. The dominance of Japanese platform manufacturers has allowed Japanese video games, as well as its comics and animated films, to exert a significant influence upon global pop culture. However, with the entry of Miscrosoft into the platform market, the monopolistic control of the Japanese industry over hardware is disappearing. It remains to be seen whether the contents of video-game software will retain the influence from Japanese culture or enter a new era characterized more strongly by Anglo-Saxon cultural content.

Acknowledgments

The original, longer version of this article appeared as "Hardware gimmick or cultural innovation? Technological, cultural, and social foundations of the Japanese video game industry," in *Research Policy*, 32 (3): 423–44, 2003. We would like to thank Elsevier Science for permisson to reproduce part of the article.

Notes

1 The primary focus of academic research available in the Japanese language has been either on corporate strategies (Fujikawa, 1999; Fujita, 1999a, 1999b; Sunagawa, 1998), or on the inter-firm relationships within the industry (Shintaku, 2000; Shintaku and Ikuine, 1999, 2001; Tanaka and Shintaku, 2001; Yanagawa and Kuwayama, 2000).
2 Despite their phenomenal popularity and profound influence on popular culture, cartoons and animated films have long been considered a societal vice by Japan's parents and educational establishments. A dramatic shift in their views has been observed only recently. Asahi.com reported on 28 September, 2000 that the Education White Paper produced by the Ministry of Education focuses on the role of culture in education, which includes cartoons and animation as an artistic sub-field, and for the first time recognized their role as art form and a form of expression that illuminates contemporary social trends.
3 The great majority of animation TV series broadcasted are domestically produced. In 1999, only 296 out of 2,553 animation programs broadcast in Japan were foreign programs (Dentsu Soken, 2001). Japan is a large exporter of TV animation programs as well; in 1992, 58.3 per cent of TV programs exported were animations (Kawatake and Hara, 1994).
4 While game-software publishers are dominated by small firms, firms in the cartoons and animation film industry are generally even smaller. In 1999, there were 235 firms in business, employing 3,720 persons (Dentsu Soken, 2001).
5 Interviews by authors with Jun-ichi Kawamura, Chief Producer, Namco, 12 April 2001 and with Komei Harada, middleware programmer at 3DO, 18 August , 2001.
6 For a detailed analysis of American comic culture and animated films, see Bendazzi (1994) and Wright (2001).
7 In the mid-1980s Nintendo was sued by Warner, which claimed that "Donkey Kong" violated copyright laws for its similarities to *King Kong*. The lawsuit ended in Nintendo's favor, when it became clear that Warner never officially copyrighted *King Kong* (Sheff and Eddy, 1999).

Bibliography

Aida, Y. and Ohgaki, A. (1997) *Shin Denshi Rikkoku: video geimu kyofu no kobo*, Tokyo: Nihon Hoso Shuppan Kyokai.

Anchordoguy, M. (2000) "Japan's software industry: a failure of institutions?," *Research Policy*, 29: 391–408.

Bendazzi, G. (1994) *Cartoons: One Hundred Years of Cinema Animation*, London: John Libbey.

Benjamin, W. (1936) "The work of art in the age of mechanical reproduction," trans. Zohn, H., *Illuminations*, 1969 edn, New York: Schocken Books.

Bryman, A. (1997) "Animating the pioneer versus late entrant debate: an historical case study," *Journal of Management Studies*, 34: 415–38.

Coase, R. (1937) "The nature of the firm," *Economica*, 4: 386–405.

Clifford, L. (2001) "This game's not over yet: Xbox vs. PlayStation: who cares? The real winners are videogame makers," *Fortune* (9 July): 164.

Computer Entertainment Software Association (CESA) (2000) *CESA Game White Paper*, Tokyo: CESA.

Cottrell, T. (1994) "Fragmented standards and the development of Japan's microcomputer software industry," *Research Policy*, 23: 143–74.

Crosby, O. (2000) "Working so others can play: jobs in video game development," *Occupational Outlook Quarterly*, 44: 3–13.

Dai-X Shuppan (2000) *Entateinmento Gyokai Shushoku 2002: gemu*, Tokyo: Dai-X Shuppan.

Dentsu Soken (1999) *Joho Media Hakusho 1999*, Tokyo: Dentsu.

—— (2001) *Joho Media Hakusho 2001*, Tokyo: Dentsu.

Department for Culture, Media and Sport (1998) *Creative Industries: Mapping Document*, London: Department for Culture, Media and Sport.

Economist, The (1995) "Eclectic: Japanese manga," 16 December: 116–18.

—— (2000) "Play to win: games and the Internet go well together. Survey of E-Entertainment," 7 October: 12–16.

Fujikawa, Y. (1999) "Sofuto kaihatsu o suishin suru dainamizumu no gensen," in Shimaguchi, M., Takeuchi, H., Katahira, H. and Ishii, J. (eds), *Seihin kaihatsu kakushin*, Tokyo: Yuhikaku, 363–87.

Fujita, N. (1999a) "Famicon tojo mae no nihon bideo gemu sangyo," *Keizai Ronso*, 163: 59–76.

—— (1999b) "Famicon kaihatsu to bideo gemu saygo keisei katei no sogoteki kousatsu," *Keizai-Ronso*, 163: 69–86.

Gertler, M. S. and DiGiovanna, S. (1997) "In search of the new social economy: collaborative relations between users and producers of advanced manufacturing technologies," *Environment and Planning A*, 29: 1585–602.

Granovetter, M. (1985) "Economic action and social structure: the problem of embeddedness," *American Journal of Sociology*, 91: 480–510.

Greenfield, P. M. (1984) *Mind and Media: The Effects of Television, Video Games, and Computers*, Cambridge: Harvard University Press.

Grigsby, M. (1998) "Sailormoon: manga (cartoons) and anime (cartoon) superheroine meets Barbie: global entertainment commodity comes to the United States," *Journal of Popular Culture*, 32: 59–80.

Hall, P. (1998) *Cities in Civilization: Culture, Innovation and Urban Order*, London: Weidenfeld & Nicolson.

—— (2000) "Creative cities and economic development," *Urban Studies*, 37 (4): 639–49.

IDSA (Interactive Digital Software Association) (2000) *1999 State of the Industry Report*: http://www.idsa.com.

—— (2001) *Economic Impacts of Video Games*: http://www.idsa.com.

Jenkins, H. (2000) "Art form for the digital age: video games shape our culture. It's time we took them seriously," *Technology Review*, 103: 117–20.

Kawatake, K. and Hara, Y. (1994) "Nihon wo chushin to suru terebi bangumi no ryutsu jokyo," *Hoso Kenkyu to Chosa* (November): 2–17.

Kinder, M. (1991) *Playing with Power in Movies, Television, and Video Games: From Muppet Babies to Teenage Mutant Ninja Turtles*, Berkeley: University of California Press.

Lent, J. A. (1995) "Cartoons in East Asian countries: a contemporary survey," *Journal of Popular Culture*, 29: 185–98.

Llewelyn-Davies, UCL Bartlett School of Planning and Comedia (1996) *Four World Cities: A Comparative Study of London, Paris, New York and Tokyo*, London: Llewelyn-Davies.

Loftus, G. R. and Loftus, E. F. (1983) *Mind at Play: The Psychology of Video Games*, New York: Basic Books.

Mikuriya, D. (1999) *Gemu yori 100 bai omoshiroi gemu gyokai no nazo*, Tokyo: Eru Shuppansha.

Ministry of Industry, Technology and Economy (METI) (2001a) *Dejitaru kontentsu hakusho 2001*, Tokyo: Dejitaru Kontentsu Kyokai.

—— (2001b) "Joho sabisugyo," *Tokutei sabusu sangyo doutai toukei geppou: heisei 13 nen 2 gatu bun*; available online: <http://www.meti.go.jp/ statistics/downloadfiles/ hv41200j.xls>

Monopolies and Mergers Commission (1995) *Video Games: A Report on the Supply of Video Games in the U.K.*, London: HMSO.

Nelson, R. R. and Winter, S. G. (1982) *An Evolutionary Theory of Economic Change*, Cambridge, MA: Belknap Press.

Nikkei B. P. (1999) *Anime Bijinesu ga Kawaru*, Tokyo: Nikkei BP.

Oshita, E. (2001) *Enix no hiyaku: Jitsuroku gemu gyokai sengokushi*, Tokyo: Shouin.

Osborn, R. and Hagerdoorn, J. (1997) "The institutionalization and evolutionary dynamics of interorganisational alliances and networks," *Academy of Management Journal*, 40 (2): 261–78.

Piore, M. J. and Sabel, C. F. (1984) *The Second Industrial Divide: Possibilities for Prosperity*, New York: Basic Books.

Pratt, A. C. (1997) "The cultural industries production system: a case study of employment change in Britain, 1984–91," *Environment and Planning A*, 29: 1953–74.

Reinmoeller, P. (1999) "Nihon no kontentsu no kokusaika," in Shimaguchi, M., Takeuchi, H., Katahira, H. and Ishii, J. (eds), *Seihin kaihatsu kakushin*, Tokyo: Yuhikaku, 388–414.

Schrodt, F. L. (1983) *Manga! Manga! The World of Japanese Cartoons*, Tokyo: Kodansha International.

Scott, A. J. (1997) "The cultural economy of cities," *International Journal of Urban and Regional Research*, 21: 323–39.

—— (1998) "Multimedia and digital visual effects: an emerging local labor market," *Monthly Labor Review* (March): 30–8.

Sheff, D. and Eddy, A. (1999) *Game Over: press start to continue*, New York: Vintage Books.

Shintaku, J. (2000) "Gemu sofuto kaihatsu ni okeru kaihatsusha manejimento to kigyo seika ni kansuru kenkyu," September; available online: <http://www.e.u-tokyo.ac.jp/ ~shintaku/TVGAME/ index_j.html>

Shintaku, J. and Ikuine, F. (1999) "Katei-yo gemu sofuto ni okeru kaihatsu senryaku no hikaku: kaihatsusha kakaekomi senryaku to gaibu seisakusha katsuyo senryaku," ITME Discussion Paper No. 22, University of Tokyo, March.

—— (2001) "Amerika ni okeru katei-yo gemu sofuto no shijo to kigyo senryaku: genjo hokoku to nichibei hikaku," ITME Discussion Paper No. 47, University of Tokyo, March.

Sunagawa, K. (1998) "Nihon gemu sangyo ni miru kigyosha katsudo no keiki to gijutsu senryaku: Sega to Namco ni okeru sofutouea kaihatsu soshiki no keisei," *Keieishigaku*, 32: 1–27.

Takano, M. (1994/1995) "Famikon kaihatsu monogatari," *Nikkei Electronics*, various issues.

Takeda, T. (1999) *Nintendo no Hosoku, Digital Entertainment 2001*, Tokyo: Zest.

Tanaka, T. and Shintaku, J. (2001) "Gemu sofuto sangyo ni okeru kigyo soshiki to seika: Kakaekomi-gata to gaibukatsuyo-gata no hikaku," ITME Discussion Paper No. 85, University of Tokyo, July.

Ueda, S. (1998) *Yokuwakaru Gemu Gyokai*, Tokyo: Nippon Jitsugyo Shuppansha.

United States Congress (1994) *Violence in Video Games*, Hearing before the Subcommittee on Telecommunications and Finance of the Committee on Energy and Commerce, House of Representatives, One Hundred Third Congress, Washington, DC: U.S. GPO.

von Hippel, E. A. (1988) *The Sources of Innovation*, New York: Oxford University Press.

Williamson, O. E. (1975) *Markets and Hierarchies: Analysis and Antitrust Implications*, New York: Free Press.

Wolf, M. J. P. (2000) *Abstracting Reality*, Lanham: University Press of America.

Wright, B. W. (2001) *Comic Book Nation: The Transformation of Youth Culture in America*, Baltimore and London: Johns Hopkins University Press.

Yanagawa, N. and Kuwayama, N. (2000) "Kateiyo bideo gemu sangyo no keizai bunseki: atarashii kigyo ketsugo no shiten," ITME Discussion Paper No. 45. University of Tokyo.

8 Making a living in London's small-scale creative sector

Angela McRobbie

This chapter offers a schematic framework for examining questions pertaining to the actual practices of making a living in specific areas (creative micro-enterprises) of London's growing cultural sector. It comprises six short sections presented in the form of abbreviated commentaries on factors which will inform future work in this field. The chapter also draws on two case studies taken from the ongoing research. Each of these throw light, in different ways, on the space–time dynamics of "art work" as it is developing as a cluster in east and south-east London. The aim overall is to present an account which considers the interface between macro-level understandings of the global city and micro-level analysis of working lives. But this focus on London does raise various issues. In the context of the growing interest over the last decade within the social sciences and cultural studies in the "new culture industries," it is often claimed that London has attracted too much attention already. More widely it has been noted that there are perhaps problems in relation to looking primarily towards the global cities of the West like London as archetypal spaces of flows, as embodiments of the network society, and also as sites of cultural innovation. How do we understand their exceptionality, their disembedded or "lifted out" status (Giddens, 1991)? And what kind of standing do policy decisions have when designed exclusively for a global city like London? Soja has referred to "Londonised myopia," suggesting that such attention can be at the expense of analyzing more ordinary cities, let us say Leeds or Liverpool (Soja, 2002). Notwithstanding the complex issues raised in regard to how government develops "customized" policies designed to reap the rewards of London's global city status, examination of this "exceptional" environment does allow us to explore how this new cultural economy actually works. The U.K. government is now pinning many of its hopes on the success and longevity of this sector, as is apparent in the volumes of documents, White Papers and Green Papers, reports and glossy brochures published in the last five years. For this reason, how the labor markets of the creative sector operate at ground level, and the kinds of livings that are being made in London's creative industries, are surely areas worthy of examination.

A recent document on London's creative sector produced by the Greater London Assembly provides some useful information in relation to these questions (GLA Economics, 2003). First that there has been substantial growth in the city's culture industries in the five-year period from 1995–2000, making this London's third largest sector, with a workforce of almost 0.5 million people. The scale of the capital's domination in the creative industries is also marked, with 110,200 new jobs created in London alone over this period and a further 86,000 in the South East in contrast to job losses in other parts of the country. What the document does not highlight, however, are the more telling aspects of the "growth narrative"; for example, with the exception of radio and television the average size of cultural-sector enterprises is less than 25 persons, so these are very small organizations. Likewise the authors do not highlight the fact that while 25 percent of London's working population have no qualifications at all, 41 percent of persons in the creative industries are "professional" and 31 percent have university degrees. This too gives to the culture industries a uniqueness and distinctiveness, comprising small outlets with very well-qualified persons. There are already a range of features in place which make London particularly attractive to well-qualified arts graduates and "young creatives." There is of course its prominence as the center for the entire U.K. media and communications complex (press, magazines, television, fashion, film and music) but there are also other features ("history, agglomeration and locational specificity"), which mean that it would be more surprising if there was not a lively fringe of small scale semi-entrepreneurial creative activities wrapping themselves around the bigger organizations and companies (Scott, 2000).

A closer look at London's cultural sector from the perspective of working lives allows consideration of important but overlooked issues. If Castells (2002) is right when he claims that workers are "increasingly left to themselves" without any organizational support or representation, we might ask how in the expanding cultural sector people carry out their creative "projects," what are the kinds of working practices which emerge, what are the distinctive "work cultures" (Sassen, 2002)? Where this kind of documentation has been undertaken the focus has been more on process and on the organization of creative labor and less on the space of such activity (e.g. Negus, 1996; McRobbie, 1998, Patterson, 2001; Nixon, 2003). Not a great deal is known about the actual formation of a whole web or network of small ventures. Exactly how a gravitational pull exerts itself drawing people into areas and how in turn this impacts on the existing community has been of course the key issue in gentrification debates (Zukin, 1988). But even here the concern is less on day-to-day working practices and on the logistics of making a living than on the consequences for the wider community and the real estate companies for whom profit margins rise exponentially with the presence of the artists.

By combining detailed attention to working practices with an equally developed sense of space and by interviewing and visiting on site a range of these people it might also be possible to get a sense of how they operate as subjects within the "space of flows," how they also flow through the capital, its institutions, organizations and neighborhoods, and how in turn the city is opened up, transformed by such movements so that its fluidity is understood in terms of this specific kind of labor and not just capital (Castells, 1996). Do its spaces or neighborhoods perhaps undergo change in terms of both de-territorialization and "re-territorialization of local sub-cultures" (Sassen, 2000, 2002)? It might also be possible to define this workforce in transnational terms. How does the work they do connect with where they have come from? Do these other locations figure in the working practices they carry out in London? Many of these workers are based in London but not necessarily producing for the U.K. market, many if not all of their clients or outlets might be from anywhere in the world. In their working practices they might be passing through London, using it as a site for gathering more skills and expertise before moving on. To what extent then is their attachment temporary or transitional; what is the creative labor flow in and out of the capital? With all of these questions in mind let me surmise that we might expect London's small-scale creative sector to comprise in the near future of a cultural workforce of more mobile and "transient" individuals on the basis that as jobs and contracts are increasingly casualized and internationalized, creative freelancers find themselves even more nomadic. London becomes a temporary site of intense cultural activity, requiring enormous investment for newcomers (in terms of cost of living) but providing access to valuable learning experiences and skills not found elsewhere. Over time the small-scale freelance cultural workforce will fragment and disperse, and more hopeful newcomers will arrive, all of which in turn suggest that networks will be formed and dissolved more rapidly and according to the duration of "projects." Relation to space and attachments to neighborhood will be attenuated and instrumental, and subject to the logic of this speeded up de-territorialized cultural economy.

Pleasure and pain

There is a substantial literature on the relation between the global city as a space of culture, and the way in which cities like London, New York and Paris, and the kind of urban policies which have been pursued in recent years, foster transformations in the world of work, including the decline of the public sector and the growth of the service sector, the new media and, of course, enterprise culture (Sassen, 1991; Scott, 2000). But even when creative work is the focus of close attention what is frequently missing is the importance of the pleasure–pain axis as a shaping characteristic of this kind of work (McRobbie, 2002). The role of affect in creative labor, and the normative expectation of the pain of insecurity, precariousness and even failure, skewers any comparison with more standard work or employment. Professed "pleasure in work," indeed passionate attachment to something called "my own work," where there is the possibility of the maximization of self-expressiveness, provides a compelling status justification (and also a disciplinary mechanism) for tolerating not just uncertainty and self-exploitation but also for staying (unprofitably) within the creative sector and not abandoning it altogether (McRobbie, 1998). More recently this recognition of work in a post-industrial context offering a kind of inherent reward, something in addition to the idea of self fulfillment in "normal work," is found in the concept of "immaterial labor," i.e. the product of the social tendency towards collapsing boundaries here manifest in the indistinguishable lines between work and play (Lazzarato, 1996). There are obvious difficulties with such a proposal, not least that the blurring of work and play might also in practice eliminate the existence of play-time altogether (see later). But, as Terranova argues, the quintessential prototype for this kind of over-working (through the night) on the basis of self-expressive output (e.g. imaginative website) is found in the gift economy dynamics of internet working (Terranova, 2000, and see also Ross, 2003). There are some points for further consideration here. Does the excessive emphasis on talent, uniqueness and creativity in cultural work act as a disconnect factor in relation to other kinds of more normal work for people with similar qualifications, for example teaching or publishing? Does this in a sense isolate and further individualize those trying to make a living in the talent-led economy? And when their numbers swell (as they have been doing in the last few years) what are the prospects of "association"? Will they ever join forces to collectively challenge the conditions which give rise in the longer term to (well-documented) assorted pains (including mental and physical illness, sporadic earnings, loss of confidence and so on)?

Across the terrain which includes graphic designers, new media workers, self-employed arts administrators and curators, fashion designers, stylists, and many other creative actors, there is a complete absence of labor organization, with the effect that a now expanded labor market in the cultural sector takes the lead not from unionized actors and actresses (stage, screen and TV), television workers or journalists but instead from a blend of the bohemian individualism of artists and the business ethos of the commercial art director. The small-scale independent company (of perhaps two or three people) and the non-organized casualized freelancer come to represent the dominant units of cultural production. In recent years with the exponential growth of freelance work replacing contract work, with the end of the "closed shop" in television and in print journalism, with the streamlining of big organizations, and with a vast population of new entrants wishing to join this labor market, union organization along traditional lines is either seen as irrelevant or simply by-passed. In any case labor relations less frequently (indeed rarely) comprise a standard contract between employers and employees. The interface of power becomes both more fluid and opaque. As Patterson explains, in television it is the commissioning editors who yield enormous power and the individuals on the other side of the fence whose future livelihoods depend on a combination of reputation and speedy access to a network (Patterson, 2001). But the displacement of power in this kind of work, away from the conventional oppositions of manager and workforce and the absence of union representation, only makes antagonisms acutely felt but undirected and often inner-directed (Bauman, 2000). Inequities, injustices and malpractices are widely recognized, almost normative, but rarely confronted. The demands of the network (pubs, clubs, hanging out) are frequently such that various categories of persons (e.g. single mothers) or those without social capital (without a university degree) are precluded, or only gain access with difficulty (McRobbie, 2003). Thus there are new barriers to entrance to replace the old closed shop. There is then a single key factor which provides a framework for understanding which is that this is largely self-organized, entrepreneurial activity such that (part-time) employment law only partially or spasmodically or periodically applies.

Space of flows

Organizational fixity gives way to fluid, flexible and placeless work while network sociality produces its own mysterious, transient yet intimate geographies. These clubs and bars and other locations are predicated on inter-personal exchange and bodily presence. For people seeking this kind of work there are often multiple levels of subcultural capital required to navigate one's way in the direction of a job or a project (Thornton, 1996). And with such high degrees of uncertainty the workforce must always be in a "state of readiness" in that the next contract or project might be bigger or better or lead to greater things (Lash and Urry, 1994). As Patterson points out, in television, starting on one job is the point at which it becomes necessary to start thinking about the next job. If the current job is a temporary resting place from which the search for the next job can commence, the nature of the interactions and the exchanges among those working alongside each other will surely reflect some of these tensions. This kind of semi-detached relation to the work at hand also denotes the absence of permanence and durability which Sennett argues were once the features which made work (and the workplace) and life meaningful (Sennett, 1998). Where the work itself is also carried out in some indeterminate space, a hot desk or a temporary office, even a local coffee bar, any prospect of stability or security is likely to give way to feelings of impermanence and insecurity. There is, as it were, no floor, no set rules, no guaranteed pathways. We might surmise that in such an unstructured and individualized work culture, new kinds of groups, affiliations or partnerships as well as relations of dependence and obligation (i.e. of a non-standard, non-contractual type) might appear, defined according to the spatial relations within which the cultural activity takes place.

Wittel defines network sociality as a form of social bonding which is "based on individualization, and deeply embedded in technology; it is informational, ephemeral but intense and is characterized by an assimilation of work and play" (Wittel, 2001: 71), but as he notes there is little ethnographic or observational research on these brittle and perhaps shallow exchanges. It is my intention here to present work-in-progress which strategically locates my own place of work, Goldsmiths College, as the hub of such a network. Its "hub" status derives partly from its historical reputation in the field of training visual artists. Its geographical location in an area of deprivation with a high-density population with low skills and few qualifications means that this poverty zone provides a wider setting for the kind of network that comes into being. This is in contrast, let us say, to one that might exist around other art schools in more affluent, or simply busy city-center settings. The college is also home to a range of academics ("radical professionals") whose work spans cultural and media studies, anthropology, art history, sociology and urban studies. This in itself is generative of a hub insofar as seminars, exhibitions and other events bring various urban activists, youth and community workers and public-sector arts administrators into the corridors and seminar rooms of the college.

Each of the two case studies which follow are based in art worlds. Each individual or group made some quite unsolicited contact or else was present at a meeting or event, or took part in some initiative which connected indirectly with the college. This coming forward, or presenting oneself as available to be involved in a research project as a respondent, suggests the existence of a quasi-network and an awareness on the part of the independent cultural producer that there is some as yet unrealized potential in the interactions between this kind of academic institution, its "experts" and themselves, the entrepreneurs. Or perhaps academics actually offer something of a degree of permanence and security in these encounters, belonging to a time of "narrative sociality" (Sennett, 1998). The two case studies reported here are a small proportion of the (self-presenting) respondents.[1] It was a condition of participation that respondents have some indirect (or "flowing," passing-through) relation to the college without being enrolled as a student. The self-presenting relation (in preference to, say, the use of volunteers) is in itself a distinctive feature of this kind of network analysis. It shows for example how the academic institution is sought out as a (public-sector) haven from the more commercial aggressive interfaces of pitching, bidding and late-night networking for contracts. The flow of knowledge, expertise, "theory" and information within the confines of the academy is also more open, inexpensive, and less constrained by the fear of "theft of ideas" so common in the creative world. The college, as a public-funded organization, has an ethos of collaboration, partnership and in recent years a keen interest in the small-scale arts enterprises (or incubator businesses as they have come to be known) springing up on its doorstep. The direct approaches to myself by these cultural entrepreneurs can also be seen as sharp indicators of their own reflexivity, awareness of the need to test out their own practices in the presence of sociologists and cultural theorists. With a deliberately open approach to methodology led by a sense of the space of flows, it was useful to concentrate on the various nodes of activity taking place at a satellite distance from the college hub in the form of *events, encounters and initiatives.*

"Assembly"

Between the 5th and the 31st of October 2000, 174 artists exhibited work together in two old school buildings due for demolition in East London. With sponsorship from David Bowie (www.bowieart.com) and others, graduates of Goldsmiths College, the Royal College of Art and Chelsea College took over the derelict spaces and under the title "Assembly" staged a show that attracted a lot of press attention and very favorable reviews. The catalogue listed all the contributors with either a phone number, an address or email address (and in some cases all three). Contact was made by email in July and August 2001, first to find out if the artist would be willing to take part in an email exchange (given the title "Making a Living as a Visual Artist"). Of the 137 email addresses provided, about 15 were returned as no longer existing; there were also about 25 politely negative replies from others now living outside the U.K. And in the end 41 agreed to take part, returning the questionnaires during an eight-week period. There was then a three-stage approach: first a questionnaire, then a request to keep an email diary, then a studio visit for a fuller interview and discussion. Thirty full questionnaires were returned completed, while the remaining 11 were semi-completed (i.e. only a few of the questions answered and so discarded). In total seven diaries were kept, and within the timescale of the project only three studio visits were made.

This small research initiative provided valuable insight into the daily working practices and the economies of young artists in London. The age ranges were all between 25 and 36, with one former GP turned artist aged 44, and two others were 39. Of the 30 respondents there were equal numbers of male and female, and a wide range of nationalities, including one Pole, one Bosnian, one Colombian, two French, one American, one German, one Austrian, two Danes, one South African and one Swiss, and the rest were British. The majority lived and worked in East London, only five in Lewisham, and a few others in more far-flung parts of London, but generally having a studio in East London. The most common provided postcodes were E1, E3 and E11. Almost all of them lived in shared flats, paying around £300 a month in rent, while more than half had studios with monthly rent varying from unusually cheap (£60) to average £100 per month with a handful a good deal more expensive at £200. Those who did not have studios were able to work at home. These were all highly qualified young people, many with up to three degrees. The questionnaires also showed an exceptionally high level of professional commitment. The art work took priority but was funded through a wide range of jobs; many respondents had at least two jobs at the same time, and these were taken on as a means of supporting their "own work." Taking all the collected data together (including the diaries) a number of themes emerge:

- The three art colleges provided a point of contact for bringing together as a "one-off project" an exhibition whose uniqueness was its scale, ambition and enterprise in terms of space acquisition, sponsorship and publicity. The positive reviews in the press provided strong material for the CV and portfolio; however, the network existed primarily in this instance as event.
- The transnational make-up of the respondents was also marked, the majority of whom were non-U.K. nationals, "passing through" London in career terms, and already living elsewhere eight months later.[2] Those who were most responsive in relation to the research (especially those who kept a

diary) remained in London, had studios or lived in East London and were persistently self-presenting by sending many email and hard-copy invitations to all subsequent shows. Indeed, the reciprocity factor from the artist's point of view with a research project like this hinges on contacts with interested academics who might also become reviewers, or write a catalogue essay. Maintaining contact with an institution like Goldsmiths (where part-time tutoring provides a stable income for many arts and media practitioners) and demonstrating an instinctual "network sensibility" which privileges "keeping in touch" as an investment strategy, by means of an email address which carries "academic" rather than "business" values, is how this kind of network actually works.

- In relation to making a living within the space of flows, with particular reference to London, and in this case South East London, there were exceptionally high degrees of mobility at local and translocal levels. The respondents converged in East London for studio space and for living accommodation but none of them indicated any special attachment to "community" or "neighborhood" nor was their paid work in any way related to the area and its social or public services.
- The level of activity was quite frenetic, with each respondent working from dawn to dusk up to seven days a week, criss-crossing London's many neighborhoods throughout the day for the purposes of paid jobs and also art work. None had children, though there was mention of partners. However, relatively little time was given over to domestic life, or indeed to leisure, with cooking and eating with friends or partner or a trip to the cinema, usually also work-related, the only periods of time off. An overwhelmingly organized and highly structured schedule put paid to the chaotic ideal of bohemian lifestyle – these were clearly "career artists."
- Despite this heavy investment in art work, financial returns were minimal. Those who had sold work over the last year came from the Royal College of Art (known for both its business acumen and for painting and sculpture, both of which are sellable, unlike video or installation work). Of the tiny number of artists who had sold work, all but one earned less than £1,000 from sales, the exception being Gordon Cheung, who had earned £10,000 from sales or leases. The artists spent a lot of time applying for grants, attending seminars with artists' advisors, and hoping to find some means of supporting the cost of their work and studio space, while at the same time still owing substantial amounts for student loans, business loans and credit cards. They lived from the assortment of jobs which took up a great deal of time with the result that art work was often squeezed into tiny corners of the day or evening. One woman wrote in her diary that she was so tired when she eventually got to her studio, on a Sunday afternoon, that she fell asleep. Another woman taught up to 20 hours a week of aerobics classes, and another had the same numbers of hours of TEFL classes, and one worked a 30-hour week in the Tate Modern bookshop. Another respondent, fully active in chasing commissions and making grant applications as

well as playing a key organizational role in the "Assembly" show, reported that her part-time job took up 42 hours a week.

To sum up, there were three different ways of earning a living to support the art work. By far the most popular was mainstream art-college teaching for the contacts it brought, as well as access to materials and to a library. However, this was highly sought work and often hard to come by. Next were art-related jobs, which again had some advantages in terms of contacts and network, access to gallery opportunities, technical equipment and a chain of other freelance jobs. These included arts handler, i.e. ferrying art works in vans across the city ("very good for contacts"), visual merchandiser (window dressing), graphic designer, photographer's assistant, record producer, curator ("I made contacts with other artists who are already more settled down, but this has not assisted my career in the art world yet") and commercial photographer. Finally, there were non-art-related jobs, including cleaning work, teaching (aerobics, TEFL, in FE colleges) translation work, marketing and PR, sales assistant, maitre d, as well as "temping." This level of activity suggests a degree of realism; it is assumed that art work on its own is unviable, but far from this being a problem it is taken for granted so that "other paid work" is used to prop up and provide the financial underpinning for the real work which is the primary source of identity and of self-value and status. This mode of cultural production shares much in common with the novelist who writes at night after a day's work as a sub-editor or proof-reader, or the actor who dare not give up the "day job." Secondary activities support the primary art work in the hope or expectation that either the art work will eventually pay off or else that a good enough teaching job will allow the artist to abandon at least some of the other work.

What marks out this mode as distinctive in the context of London and the creative industries is not just the multi-tasking and the transnational flow of artists in and out of the capital, but the network sensibility, the know-how about useful contacts and keeping in touch, the cultural expertise in terms of keeping up to date with new art and cultural theories, with films and exhibitions, with the need to be doing "research," with planning and self-promotion and the various means of doing this, in short, with the serious business of being an artist. London provided the possibility of finding the kind of work which could maintain the art identity, Goldsmiths was clearly recognized as a hub within the artists' networking activities, and East London marked a convergence site for studio and also exhibition space. There was little time for domesticity, and living space was defined in terms of affordability; such was the speed of flows that special attachment to the spaces of neighborhood or community was rarely commented on. Despite the difficulties there was a high degree of optimism, singularity of purpose and dedication to the art work as a source of self-identity. These respondents fulfilled Rose's account of the emergence of the entrepreneur of the self: "individuals are to become, as it were, entrepreneurs of themselves, shaping their own lives through the choices they make among the forms of life available to them" (Rose, 1999: 87).

The Hales Gallery: "this place is important to us"

Paul Hedges, director of the Hales Gallery in Deptford High Street, could be described as self-presenting in that he was invited by a colleague to attend a lunch-time discussion for about ten people, where the plans for a new arts complex were being aired. His wide knowledge of the area and of the various arts community initiatives going on suggested he was an ideal candidate for the research project. About 30 minutes' walk from the college, the gallery is tucked behind possibly the only inviting looking place to eat in the area. The restaurant area combines features of middle-class and bohemian lifestyle in a way which is not off-putting to locals. The gallery itself is downstairs and in an extension along the garden area; in addition there is a large warehouse in what was the local church hall. The first part of our visit comprised a guided tour of the premises, including a close look at the many works in the warehouse as well as the work on show in the exhibition space. In sharp contrast to the usual "moneyed" image of the art dealer, Paul Hedges had a strong social investment in his work. He was a 1983 graduate of the college, someone who had grown up in the area, maintained a strong working-class identity and professed an absolute loyalty to the neighborhood: "class is still a huge issue, there are difficult barriers if there is no money in the background and no contacts in the art world." His "self-biography" was, as is the case with most creative workers, completely entangled with his career through the art world and his particular narrative was based on being a working-class man with a social conscience and also a connection with the local church, making his way through the overtly "snooty" (as he put it) London art world.

Hedges had, while still an undergraduate, curated exhibitions in his own student house. He had worked as a postman for nine years in the hope that the time off in the afternoon would allow him time to do his own work. When this was unviable he found a derelict church space in Deptford, and with the agreement of the church (to which he belonged), and with the help of friends, he gutted it and opened it as a café. He had been involved with the church on youth projects and drug-rehabilitation schemes and so knew the ropes in terms of community involvement. He raised almost £85,000 partly by presenting to the Department of Trade and Industry the idea that this would be a "business training for artists' project. He had the gallery going at the back of the café where he himself cooked for the first six years: "I was getting up at 5 a.m. and making a load of sandwiches and then by 3 in the afternoon I got out of the kitchen to do the paperwork for the gallery." By then another graduate from music college offered to take over the cooking, and Paul and his business partner were able to devote all their time to the gallery.

The first show they put on was for Jake and Dinos Chapman, and they also were one of the first galleries to show the work of Mike Nelson. Hedges says he is "evangelical about what we do." There is "social inclusion which I believe in" and ordinary people are not put off by the gallery. Despite great pressure to move to a West End location (where the well-known dealers have their galleries), Hedges insists "this place is important to us." With 15 artists (of whom some are very well known) the Hales Gallery is a landmark in South East London. Hedges is critical of the younger generation of graduates. He claims they have none of the collaborative energy and do-it-yourself ethos: "They believe you can have a career as soon as you leave art school, selling pieces at £4,000–5,000 a time. They also think art galleries are run by crooks who will rob them." Hedges' working week now consists of keeping in touch with his artists, visiting art fairs across the world, managing the café-restaurant, doing occasional teaching slots, and also involvement in the various creative-sector initiatives in Lewisham. From art graduate with a working-class (and possibly punk?) do-it-yourself ethos, and church involvement, he has become a busy art player while retaining fixity and loyalty within the space of flows. He thus participates in what Sassen describes as "re territorialization of local sub-cultures." Hedges was critical of Goldsmiths-as-hub on the grounds of its being insufficiently entrepreneurial: "Goldsmiths doesn't know what its got. There is an enormous number of people from there who still live round here. They use our little café as the centre for an informal network – they are all Goldsmiths ex-students."

"Very little painting sales, no commissions, not signed to a gallery, no dealer." Response to email questionnaire

This "life-biography" (as Beck would call it) shows how the entrepreneurial mode dominates the career in art (Beck, 1992). Both Hedges and the "Assembly" artists are highly self-reliant. Self-organization requires connection with an art-oriented network, including a point of contact with either the public sector for teaching or in Hedges' case with local government and community for its funding opportunities. These people are all avid multi-taskers moving around the city and beyond at great speed. (Few references were made to other parts of the U.K., confirming London's disembedded or lifted-out status (Giddens, 1991). Highly disciplined and hard working, the respondents reported on endless self-monitoring of their own performances in relation to the many plans and projects they have running simultaneously. This self-reflexivity is so part of the course that the questionnaires, diaries and interviews were set upon with relish. These were also small-scale micro-economies relying on huge investments of time, mental and physical effort and there were real costs in terms of leisure, personal and family life. Hedges presented himself as successful, while all the respondents in the "Assembly" fulfilled the traditional image of the "struggling artist." London's extended macro-economy, its public institutions, its private sector and its service sector in particular provided the younger artists with at least the possibility of using other forms of part-time work to subsidize their "own work." With a working-class identity rooted in neighborhood ties, Hedges had created for himself an art career made possible by the opportunities and networks he could tap into in his own local environment and from his time at Goldsmiths, a college known for encouraging working-class students.

London's global city labor market has permitted a series of micro-enterprises of art and art-related activity whose defining features are low (or no) capital returns, but which generate high-cultural capital (Bourdieu, 1984, 1993), the value of which is exchangeable in more extended creative labor markets; this cultural capital also bestows status on participants, which compensates for the precarious life and the expense of the investment. We might then ask, is this a pattern we might expect to find emerging in other similar urban centers? We might also inquire as to the longer-term outcomes for such significant numbers of artists and those working alongside them or supporting them? The Creative Lewisham study suggests that there are more than 500 arts or creative businesses within the area, but given how recent this growth is and how few detailed studies there have been of this kind of activity (considered over the duration of, for example, a lifetime's work), the key feature has then got to be the degree of risk or uncertainty (Landry, 2001). These fully flexible units might mutate and transform and develop according to any number of variables. There is no model – these are "permanently transitional" pathways.

It is difficult also to locate the power relations at work; we might imagine them to be those which exist in and through the mechanisms of self-regulation. These are deeply inculcated at a corporeal level; they are the degrees of self-exhortation – to make it, to do better, to talk to the right people, to achieve success. Endless self-disciplining, the burden of self-assessment and the handling of self-promotion, as well as the privatization of disappointment and the internalization of grievances, suggest power to be embedded in the sometimes surely torturous practices based on the normative requirement to be motivated, to keep in touch, to make the effort, to take responsibility for personal success and the many other daily routines of self-maintenance. For the artists in the "Assembly" show these crowd out other possibilities and indeed negate by their sheer forcefulness ideas which run counter to these neo-liberal notions of endless self-invention. Gone are the kinds of radical and collaborative actions of the past which drew artists closer to marginalized or disadvantaged groups; gone too are artist community initiatives – there were no signs in the questionnaires of art connecting with social work, or with work with children, the mentally ill, or the elderly. They are radically disconnected and dislocated from "community." Even though, as both a social and cultural entrepreneur (an unusual mix), Hedges is the product of an earlier decade (i.e. the 1980s), he too is pressured to take the gallery out of the low-income area of Deptford and into the heart of the West End.

Let me conclude this chapter by suggesting that all of these participants could be described as "new subjects of cultural individualization," they are self-disciplining and self-managing, they understand themselves to be fully responsible for the choices and pathways they have followed, they enjoy this freedom and prefer to take on many tasks or projects rather than consider (even when it might be more lucrative) normal work or stable employment in an art-related field (e.g. secondary-school art teaching). London is for them all a city of network possibilities, but where there is such a burden of expectation to "make it" as an artist, the global city becomes strangely drained of life and vitality, it is de-socialized as surely as it is neo-liberalized. Or to put it another way, strong commitment to place and involvement in the neighborhoods of the global city, including "re-territorialization," are increasingly precluded or made impossible by the speeded-up economy of art working. The city becomes, not a place of living, but a shadowy backdrop for contacts, parties, events and "possibilities." The relationship to the city is tenuous, even ephemeral, not unlike that which shapes the other projects and temporary contracts. It may well be that in the near future, with the growth of a transnational cultural economy, London remains rich as a site for creative transactions and network opportunities, but becomes more locally impoverished as this increasingly nomadic workforce uses the city as a "hot desk" space for "passing through," but is unable to generate the kind of income now needed to settle in London. This in turn might well mean that the high hopes for a sustainable creative sector in the city are unviable without a policy change on the part of government. This would require a dramatic re-consideration of the value of the entrepreneurial model for the future of the creative economy.[3]

Notes

1 To date (August 2003) the full research material from the project comprises 20 extended interviews based on visits, observational analysis from events, huge quantities of promotional material from contacts and respondents received by email, the email questionnaires from "Assembly," and finally from "hub initiatives," i.e. Goldsmiths meetings and seminars.

2 It was difficult to know exactly how many of the 174 artists were temporarily in London, and what the flow rates were backwards and forwards from place of origin. 25 emails were received (mostly from South East Asia), indicating that the artists were back in the country of origin, and hence unable to take part.

3 Let me mention the case of the 35-year-old German fashion designer Vera Von Garrel, who uses London as a base for her company, and is able to take advantage of the easy means of becoming officially self-employed here (unlike in Germany). Weekdays she rents a room from a friend which doubles up as office and studio, but she lives in Munich (where rents are low), manufactures her clothes in Germany (where the quality is higher) and uses the cheap flights to travel between the two cities.

References

Bauman, Z. (2000) *Liquid Modernity*, Cambridge: Polity Press.

Beck, U. (1992) *Risk Society*, London: Sage.

Castells, M. (1996) *The Rise of the Network Society*, Oxford: Blackwell.

—— (2002) "An Introduction to the Information Age," in G. Bridge and S. Watson (eds) *The City Reader*, Oxford: Oxford University Press.

Giddens, A. (1991) *Modernity and Self Identity*, Cambridge: Polity Press.

GLA Economics (2003) *Creativity: London's Core Business*; available online: <www.london.gov.uk>

Landry, C. (2001) *Creative Lewisham*; available online: <www.lewisham.gov.uk>

Lash, S. and Urry, J. (1994) *The Economies of Signs and Space*, London: Sage.

Lazzarato, M. (1996) "Immaterial Labor," in S. Makdisi, C. Casarino and R. Karl (eds) *Marxism Beyond Marxism*, London: Routledge.
McRobbie, A. (1998) *British Fashion Design*, London: Routledge.
—— (2002) "From Holloway to Hollywood: Happiness at Work in the New Cultural Economy," in P. Du Gay and M. Pryke (eds) *Cultural Economy*, London: Sage.
—— (2003) "Club to Company," *Cultural Studies: Special Issue on "Who Needs Cultural Intermediaries?,"* 16: 516–31.
Negus, K. (1996) *Producing Pop*, London: Edward Arnold.
Nixon, S. (2003) *Advertising Cultures*, London: Sage.
Patterson, R. (2001) "Work Histories in Television," *Media, Culture and Society*, 23: 495–520.
Rose, N. (1999) *Powers of Freedom*, Cambridge: Cambridge University Press.
Ross, A. (2003) *No Collar*, New York Basic Books.
Sassen, S. (1991) *The Global City*, Princeton, NJ: Princeton University Press.
—— (2000) "New Frontiers Facing Urban Sociology at the Millennium," *British Journal of Sociology*, 51: 143–159.
—— (2002) "Extract from 'Globalisation and its Discontents'," in G. Bridge and S. Watson (eds) *The City Reader*, Oxford: Oxford University Press.
Sennett, R. (1998) *The Corrosion of Character*, New York: Norton.
Scott, A. J. (2000) *The Cultural Economy of Cities*, London: Sage.
Soja, E. (2002) "Six Discourses on the Postmetropolis," in G. Bridge and S. Watson (eds) *The City Reader*, Oxford: Oxford University Press.
Terranova, T. (2000) "Free Labor: Producing Culture for the Digital Economy," *Social Text 63*, 18: 33–57.
Thornton, S. (1996) *Club Culture*, Cambridge: Polity Press.
Wittel, A. (2001) "Toward a Network Sociality," *Theory, Culture and Society*, 18: 51–77.
Zukin, S. (1988) *Loft Living*, London: Hutchinson.

Part IV

Clustering processes in cultural industries

9 Toward a multidimensional conception of clusters

The case of the Leipzig media industry, Germany

Harald Bathelt

Introduction

This chapter deals with a subset of cultural products industries, involved in the commodification and creation of culture and the transmission of social and cultural content. These industries have been a site of intense geographical exploration (Storper and Christopherson, 1987; Lash and Urry, 1994; Crewe, 1996; Scott, 1996, 2000; Pratt, 1997; Leyshon *et al.*, 1998; Grabher, 2002; Power, 2002). Although the composition of cultural-product industries is likely to vary between different economies and societies, it is still somewhat nebulous which industries can safely be regarded as cultural product industries. There is some consensus, however, that the media and multimedia industries belong to this sector. From the perspective of economic geography, the tendency of these industries to agglomerate and develop a particular social division of labor is particularly interesting. Regional clusters of specialized, interrelated media industries are often the product of local growth processes driven by innovative start-ups (Scott, 1996; Brail and Gertler, 1999; Egan and Saxenian, 1999).

The following discussion will present the case of the Leipzig media industry in East Germany. Leipzig has a long history as a site of media industries. Its book-publishing sector, concentrated in the city's Graphisches Viertel (Graphical Quarter), dominated the German book trade from the eighteenth century until the Second World War. As of today, there is little left of this industry, due to the sectoral and political ruptures caused by the Second World War and German Reunification. However, a new-media industry has developed since the 1990s, centered around the areas of TV/film production, graphics/design and new digital media. These are the media branches which I will focus on in this chapter.

The goal of this chapter is to analyze the evolution of Leipzig's new media sector using a multidimensional cluster conception which distinguishes its development along the horizontal, vertical, institutional, external and power dimensions. The results presented are based on more than 100 interviews with media firms in the above mentioned branches which were conducted between July 2000 and August 2002. Further information about start-up processes and public policies to support this development were acquired through 20 additional interviews with local planners, policy makers and bank representatives.

The line of arguments is structured as follows. In the next section, the historical foundations of Leipzig's media sector will be sketched out. It will be shown that this sector had to undergo severe crises and periods of restructuring during the twentieth century. Building upon the concept of re-bundling, a multidimensional cluster framework is then developed which explains how clusters are created, why they grow and how they reproduce themselves. This is used in later sections to analyze the genesis of Leipzig's new-media industry cluster during the 1990s. It will be shown that firm formation and relocation processes have given rise to the vertical dimension of the cluster. I will show that this development has greatly benefited from a supplementary process of institution-building. Evidence is provided to show that the external dimension of Leipzig's media industry cluster is not very well developed, establishing a barrier for further growth. The chapter ends with some conclusions concerning the present state of the cluster and potential threats, as well as some opportunities, for long-term prosperity.

Study context: re-bundling processes in Leipzig's development path

Already in the Middle Ages, Leipzig was an important trade and service center, which had developed into a leading location for trade fairs in Europe (Schmidt, 1994; Gormsen, 1996; Grundmann, 1996). Before the Second World War, Leipzig also played a leading role in the German book-publishing industry (Schulz, 1989; Wittmann, 1999). In the 1930s, there were more than 800 publishers and booksellers with over 3,000 employees (Denzer and Grundmann, 1999). Most of the firms were located in the Graphisches Viertel, adjacent to the downtown area (Figure 9.1). The local production system at that time was that of a nineteenth-century industrial district (Boggs, 2001).

Regional ruptures and restructuring

The Second World War dramatically ruptured the evolution of the local media sector. While the Graphisches Viertel was rebuilt following the Second World War, it was integrated into the international socialist division of labor and interacted little with West Germany. Even though Leipzig was still an important location of the book-publishing industry of the G.D.R. (German Democratic Republic), it was not able to retain its status as a national and international center of these industries (Denzer and Grundmann, 1999; Gräf, 2001).

At the time of German Reunification in 1990, Leipzig's book-publishing industry, like most industries in East Germany (Kowalke, 1994; Oelke, 1997), was not well positioned for market-driven competition. In 1990, the Förderverein Medienstadt Leipzig (Development Association of the Media City Leipzig), a loosely organized public–private partnership, began to promote the re-development of the Graphisches Viertel as a site of book publishing and affiliated industries (Baier, 1992; Denzer and Grundmann, 1999; Schubert, 2000). However, this did not work. The Graphisches Viertel was neither able to grow

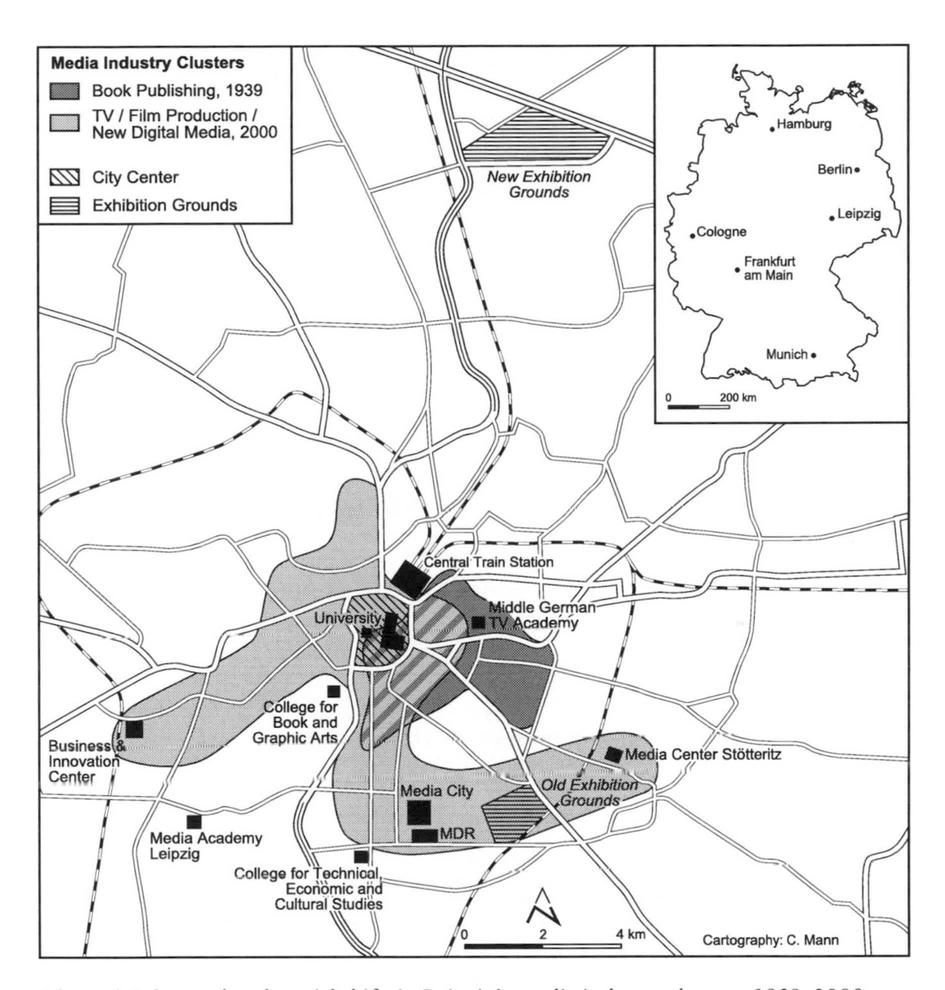

Figure 9.1 Sectoral and spatial shifts in Leipzig's media industry clusters, 1939–2000.

into a center of traditional media branches nor did it develop into a significant location of new media industries.

In the area of electronic media, such as radio, TV and film, Leipzig did not have a strong tradition (Sagurna, 1999, 2000). The G.D.R. radio, TV and film sector was mainly concentrated in East Berlin–Potsdam. Despite this, a new media industry cluster developed in the city during the 1990s around the activities of the MDR (Mitteldeutscher Rundfunk, Middle German Television and Broadcasting Service). The MDR was established in 1991 as a public broadcasting service in East Germany. It rapidly became the most important institution supporting the establishment of media branches in Leipzig.

Table 9.1 compares Leipzig to other centers of media industries in Germany. It clearly shows that Leipzig is much smaller than the major agglomerations

Table 9.1 Number of establishments and sales of media industry clusters in Germany, 1998/1999

Media branch	*Media industry clusters*					
	Germany 1998	*Munich 1998*	*Cologne 1998*	*Hamburg 1998*	*Berlin 1998*	*Leipzig 1999*
			A. Number of establishments			
Film	6,880	910	502	531	752	31
TV/radio	2,081	168	165	45	165	12
Publishing	15,183	1,475	855	744	797	182
Advertising	25,461	1,700	1,364	1,547	1,147	278
News agencies/ journalists	2,095	273	229	163	144	96
Data processing	25,725	2,032	1,141	988	1,208	133
Total	77,425	6,558	4,256	4,018	4,213	732
			B. Sales (euro millions)			
Film	7,185	3,026	499	1, 426	568	30
TV/radio	7,373	1,549	1,869	756	134	8
Publishing	38,286	2,712	1,522	4,518	3,239	283
Advertising	22,086	2,973	1,385	2,776	602	83
News agencies/ journalists	1,513	125	109	199	82	6
Data processing	27,348	4,617	2,727	961	756	134
Total	103,791	15,002	8,111	10,636	5,381	544

Sources: Deutsches Institut für Wirtschaftsforschung, 1999; Statistisches Landesamt des Freistaates Sachsen, 2001; Krätke and Scheuplein, 2001

Munich, Cologne, Hamburg and Berlin (Gräf and Matuszis, 2001; Gräf *et al.*, 2001; Sydow and Staber, 2002; Krätke, 2002) and can only be viewed as a secondary media industry center with less than 750 establishments in 1999 and total sales of 0.5 billion euros. The media industry is, however, one of the few economic sectors in Leipzig which have experienced economic growth in the post-Reunification period and has served to stabilize the local economy.

Re-bundling local assets

The sequence of crises and restructuring in Leipzig's media sector can best be explained in a model of regional development combining evolutionary processes which focus on continuity (e.g. Arthur, 1988) with unexpected ruptures and discontinuities as conceptualized in regulation theory (Lipietz, 1987; Boyer, 1988, 1990). Due to political and sectoral ruptures, ensembles of institutions that once stabilized capital accumulation can later fail. As a consequence, a regional economy's social cohesion is strained and existing transactional networks are disrupted, releasing resources for alternative uses. Suppliers and service firms which previously focussed on the needs of the dominant sector are now open to new ventures and technologies developed in other sectors and regions.

A region's recovery from crisis can be stimulated when agents re-bundle the capital at hand through interactive learning and reflexive social practices for a new round of accumulation (Bathelt and Boggs, 2003). Sometimes this becomes the basis of a new regional industry core promoting a novel development path (Storper and Walker, 1989). In order to shape a region's development so that the effects of a crisis are overcome, a new ensemble of competencies or specialized cluster of interrelated economic activities must develop. This must be anchored into the local economy by non-ubiquitous resources, found within a specialized social division of labor.

Conceptual basis: multiple dimensions and trade-offs in clusters

In this chapter, the term "cluster" is used to refer to a local or regional concentration of industrial firms and their support infrastructure which are closely interrelated through traded and untraded interdependencies (Maskell 2001; Bathelt and Taylor, 2002). This does not, of course, imply that every localized concentration of firms can be categorized as having the same growth potential and regional impact. Clusters should be analyzed along several dimensions; that is, their horizontal, vertical, institutional, external and power dimensions (Figure 9.2). Through this, different configurations of clusters can be identified according to their development stage and growth prospects (Porter, 1998, 2000; Bathelt, 2001, 2002; Malmberg and Maskell, 2002). This conception will be used to analyze the genesis and growth of Leipzig's new media industry cluster.

Competition and variation: the horizontal cluster dimension

The horizontal dimension of a cluster consists of firms producing similar products. This dimension is sometimes key to understanding why a cluster exists to

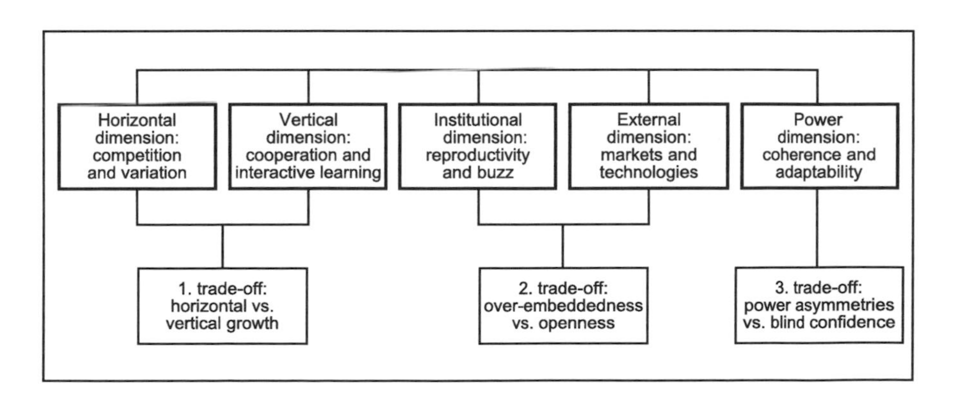

Figure 9.2 Cluster dimensions and trade-offs.

begin with (Porter, 1990, 1998; Malmberg and Maskell, 2002). Competing firms with similar competencies have little reason to cooperate with one another or engage in close interaction. Co-location provides the opportunity, however, to closely watch these competitors and compare one's own economic performance with that of others (Grabher, 2001; Maskell, 2001). Even in the absence of contact, firms know about their competitors and understand their actions since they operate under the same conditions. This creates strong incentives for product differentiation, process optimizing and cost reduction.

Although the horizontal cluster dimension in Leipzig is still underdeveloped, firms observe one another. Local competition is not very strong in the areas of TV/film production and sophisticated supplies which require substantial producer–user interaction. Primarily in the area of standardized supplies and services, one can find a larger variety of firms offering similar services. This leads to strong price competition in this segment.

Cooperation and interactive learning: the vertical cluster dimension

The vertical cluster dimension encompasses firms with complementary products and competencies which can benefit from intensive transactions within the cluster and form networks of traded interdependencies. This creates an incentive for suppliers and service firms to move closer to these customers to supply the regional market (Marshall, 1920). The firms benefit from low transportation and transaction costs, as well as economies of scale, and thus gain a competitive advantage (Scott, 1988; Krugman, 1991, 2000; Fujita *et al.*, 1999). Through this, existing clusters continue to grow, giving rise to labor-market specialization.

There are, however, more advantages to industrial clustering than those related to cost and traded interdependencies. In emphasizing localized capabilities (Maskell and Malmberg, 1999a, 1999b) and untraded interdependencies (Storper, 1995, 1997), recent studies have shown that socio-institutional settings, interfirm communication and interactive learning play a decisive role in regional innovation and growth processes (Cooke and Morgan, 1998; Lawson, 1999; Gordon and McCann, 2000; Bathelt and Jentsch, 2002). A large body of literature suggests that complex innovation processes rely heavily on supplier–producer–user interaction and corresponding learning processes (Lundvall, 1988; Gertler, 1993, 1997). As such, innovation is primarily a result of social relations and reflexive behavior and not of individualistic endeavors (Granovetter, 1985; Grabher, 1993).

In the Leipzig media-industry cluster, firm formation processes have contributed to rapid growth in the vertical cluster dimension. Meanwhile, there is a large variety of specialized suppliers and service providers which benefit from traded interdependencies. Regional producers are no longer forced to acquire most of their supplies from outside. Untraded interdependencies are also important. The development of media-related qualifications, transaction-specific investments and conventions about TV formats have enabled firms to develop common grounds for cooperation. The dynamic processes behind this will be further analyzed in the following section.

When discussing the horizontal and vertical dimensions of a cluster one has to consider that there is a trade-off between both, which could weaken the cluster's competitive basis in the long-term. Incentives for suppliers to locate within a cluster are high if the social division of labor is well-developed (Malmberg and Maskell, 2002). At the same time, the more the social division of labor is deepened, the fewer are the opportunities to make observations of competitors and secure a strong horizontal dimension. Only through continued agglomeration can the vertical division of labor and horizontal opportunities for variation be extended simultaneously.

Reproductivity and local buzz: the institutional cluster dimension

Norms, accepted rules, habits, conventions and other institutional arrangements are of great importance to enable interfirm communication and collaboration within a cluster (Amin and Thrift, 1995; Maskell, 2001; Bathelt, 2002). A joint institutional framework enables specialized users and producers to discuss and solve particular problems (Hodgson, 1988; North, 1990). Such a framework does not, however, arise spontaneously. It is created through social practices in day-to-day interactions. Joint problem-solving and experimentation lead to preliminary fixes which must be robust in order to survive the next series of interactions. These fixes are constantly updated or adjusted to new goals in the innovation process (Storper, 1997). The creation of institutions within a cluster helps to develop reasonable expectations, stimulate trust and increase stability in producer–user linkages (Granovetter, 1985; Crevoisier and Maillat, 1991; Bramanti and Ratti, 1997; Lawson and Lorenz, 1999).

Co-location and face-to-face contacts within a cluster generate further advantages related to the circulation of information and inspiration (Bathelt *et al.*, 2002). It creates a particular information and communication ecology. This local buzz is related to constant flows of information, updates of this information and intended and unanticipated learning processes which are based on the same institutional set-up (Marshall, 1927; Grabher, 2002; Owen-Smith and Powell, 2002; Storper and Venables, 2002). Actors do not have to search their environment or make particular investments to get access to this information (Bathelt *et al.*, 2002). They are automatically exposed to news reports, gossip, rumors and recommendations about technologies, markets and strategies, by just being in the cluster (Gertler, 1995; Grabher, 2002).

Such buzz is still in an infant stage in Leipzig's media industries, as firms are relatively closed and do not engage in interactive learning. Formal institutions are, however, developing within the region which support existing businesses and spur start-up processes. This will be the focus of a later section.

Markets, technologies and trans-local pipelines: the external cluster dimension

It is clear that a cluster cannot fully unfold its growth potential if its firms exclusively rely on internal markets and local knowledge. If local linkages between suppliers, producers and customers become too rigid and too exclusive, focussing on only a few local actors, this could cause problems of lock-in and pose limitations to future growth (Granovetter, 1973; Oinas, 1997; Maillat, 1998; Scott, 1998). This problem, which has also been described as over-embeddedness (Uzzi, 1997; Bathelt, 2001; Sofer and Schnell, 2002), exemplifies that there is another trade-off between socio-institutional embeddedness and openness; that is, between the internal and external dimensions. The degree of openness of a cluster has to be large enough to allow for maximum external innovation and growth impulses. Yet, at the same time, it has to be sufficiently closed to benefit from local interrelatedness and buzz.

Many studies provide evidence that partnerships with firms from other regions and countries are decisive triggers for innovation (Bathelt, 2002; Bathelt and Taylor, 2002; Tracey and Clark, 2003; Clark and Tracey, 2004). Through the establishment of trans-local pipelines (Owen-Smith and Powell, 2002), firms can tap into external knowledge pools and markets. As opposed to the local buzz, access to trans-local pipelines does not happen spontaneously. It requires conscious efforts, substantial investments and is associated with higher uncertainty (Bathelt *et al.*, 2002). Partners in pipelines are to be carefully selected and trust has to be built (Lorenz, 1999). Firms on both ends have to get to know one another, learn how to make interactive adjustments and develop an absorptive capacity (Cohen and Levinthal, 1990; Malecki, 2000). This refers to the ability to assimilate external information and apply it internally.

There is evidence which suggests that the firms in Leipzig's media-industry cluster are not open enough and have difficulties accessing other German markets regions. Their absorptive capacity is also seemingly not very strong. This serves as a barrier for further growth and could threaten the long-term prosperity of the cluster. I will return to this issue later on.

Coherence and collective adaptability: the relational power dimension

A cluster does not automatically encompass a coherent group of firms which cooperate harmoniously with one another to achieve a common set of goals. Power creates a hierarchy and rules of dominance and subordination within a cluster (Allen, 1997). Emerging power asymmetries are produced and temporarily stabilized through circuits of power (Clegg, 1989; Taylor, 1995, 2000).

To make a cluster visible as such, the internal actors, as well as those external to the cluster, must recognize the cluster as an entity which is different from its environment and act accordingly. In other words, distinct internal structures of association have to develop (Latour, 1986). If this is the case, clusters have causal power because network relations have an emergent effect (Yeung, 1994; Scott, 2001).

Social relations in a cluster are constantly being reproduced through ongoing communication, similar ways of solving problems, joint decisions about which technologies to use and the like. The power of a cluster is an expression of the potential to enroll actors in joint enterprise. Of course, it is difficult to establish coherence through social relations alone. Particular technologies serve to stabilize the interactions between firms and generate similar day-to-day experiences. Combinations of material and non-material resources and artifacts enable cluster firms to engage in social relations and keep them going (Latour, 1986; Murdoch, 1995; Jöns, 2001).

There is, of course, a danger that firms develop too much trust in a given hierarchy and become too dependent on dominant actors. This indicates another trade-off in a cluster between power asymmetries and blind confidence. The development of too much trust can cause structures of blind confidence and gullibility (Kern, 1996). This might contribute to lock firms in an inefficient technological trajectory. A certain amount of distrust with respect to traditional technologies and prominent decision-making structures in a cluster is therefore necessary to reduce the risk of collective failure.

In the case of Leipzig, too much trust is certainly not a problem. In contrast, it seems that there is not enough trust, as firms avoid overly close interaction with others. Most managers interviewed are not even aware of the already existing cluster of media activities. This shows that the cluster has not yet fully emerged as such.

Having developed a multidimensional cluster framework, the genesis and growth of Leipzig's new-media industry cluster will now be analyzed through this conceptual lens. The next section demonstrates how firm formation and relocation processes from other German regions have given rise to the cluster's vertical dimension.

Empirical results 1: the dynamics of start-up processes in Leipzig's media industries

Leipzig's media sector has grown to a substantial size. It consists of 750–1,500 media firms, depending upon which definition and data source is used (Bentele *et al.*, 1998; Statistisches Landesamt des Freistaates Sachsen, 2001). Bentele *et al.*, (2000) estimated from a postal survey that the media sector employed 25,600 permanent employees and 14,600 freelance consultants in 2000, amounting to about 16 percent of the regional labor force overall (Table 9.2). This is quite important for the regional economy, as the media sector has stabilized the otherwise shrinking labor market (Bathelt, 2001, 2002). This growth is a result of local start-up activities and the establishment of branch facilities.

Most firms in Leipzig's media sector are quite young and have few employees (Bentele *et al.*, 2000). Almost half of the survey firms were founded between 1998 and 2000 (32 of 68 firms) and about 80 percent had less than 10 employees (53 of 67 firms). Three-quarters of the media firms were established

Table 9.2 Number of employees in the Leipzig media industry cluster by branch, 1998

Media branch	*Estimated number of employees*	
	Full-time and part-time	*Freelance*
	A. Electronic and new media industries	
TV/film/radio	3,800	5,000
PR/marketing	1,500	3,700
Hardware/software	1,800	300
Data processing	700	200
Interactive media	400	100
Subtotal	8,200	9,300
	B. Print and other media industries	
Subtotal	12,800	2,500
Total 1998	21,000	11,800
Total 2000	25,600	14,600

Sources: Bentele *et al.*, 1998, 2000

by people who had lived in Leipzig throughout their entire life or by people who had studied or worked there for many years. For these firms, the question of where to locate was seemingly not a primary concern. In fact, the decision regarding the location was usually a given once the start-up decision was made. This confirms the general finding that start-up decisions are often influenced by the experience a founder has accumulated while working in a particular technological field or during his training in a specialized work area (e.g. Hayter 1997).

According to the interviews conducted, at least six types of start-up firms can be identified in Leipzig's media sector. They are important in understanding the genesis of this media industry cluster (Bathelt, 2002; Bathelt and Jentsch, 2002).

(Type 1) Euphoric local start-ups

The largest group of firms in Leipzig's media industry (one-third of the survey firms) can be classified as euphoric local start-ups. Their founders were often born in Leipzig and lived and worked here for most of their lives. They have a strong community spirit and would not have started a new venture elsewhere. These firms were founded by optimistic entrepreneurs who wanted to realize market opportunities which they had identified within their local environment.

(Type 2) Local university spin-offs

Another group of start-ups are classified as university spin-offs because the founders applied product-related know-how which they acquired during their university or college education in establishing their new venture. A number of Type 1 start-ups can also be considered as university spin-offs since their founders had studied in Leipzig.

(Type 3) Forced entrepreneurs from Leipzig

Not all of the start-ups in Leipzig are a product of visionary entrepreneurs. In some cases, the founders were forced into self-employment due to an actual or foreseeable loss of their jobs within the region. Prior to this, they had no intention of starting up their own business. The achievement of high growth rates was not a primary goal of these founders. Rather, the intent was seemingly to establish a secure economic basis for their retirement (see, also, Hinz and Ziegler, 2000). These firms were among those with the lowest sales figures in the sample. Many of the "new self-employed" studied by Thomas (2001) would fall into this category.

(Type 4) Split-offs of former state-directed firms

Especially in the area of TV/film production, some new firms were created as split-offs from former state-directed G.D.R. media firms. Rather than looking for a job in the West German media industry, the founders preferred to establish their own business with people they already knew. Leipzig was the city of choice because of the opportunities related to the establishment of the MDR.

(Type 5) MDR-related establishments from West Germany

Another identifiable group of new firms in the TV and film business (one-quarter of the survey firms) moved to Leipzig to acquire contracts from the MDR. They were established as branches of existing firms from West Germany and, more recently, firms originating from East Germany. In some cases, the MDR had asked professional film teams, technicians, cutters, reporters and other media specialists directly to establish a branch in Leipzig and offered future contracts to them.

(Type 6) Other branch establishments from West Germany

There are also a number of media firms, not directly related to the city's TV and film activities, which established branches in Leipzig. These firms are in the areas of new electronic services, graphics, public relations or marketing. They chose Leipzig as their location because of (i) the already existing media sector and its potential, (ii) the city's central location in East Germany and (iii) Leipzig's labor market and cultural amenities.

Hinz (1998) reports similar start-up motivations from his study of newly established firms in Leipzig. Two-thirds of the firms in his study were driven by positive pull factors, such as the belief in their ability to make it, their striving for economic independence or the wish to take advantage of existing opportunities. Type 4 and Type 5 firms played an important role in the development of the media sector since they brought professional expertise and specialized experience into the region which did not exist previously. Due to the combined effects of the MDR and TV/film producers, a network of specialized activities and competencies began to develop which became a trigger for Type 1 and Type 2 start-ups within the region. This created a particular dynamic. The development of the Leipzig media-industry cluster has also benefited from and is supported by a process of institution-building. This will be shown in the next section.

Empirical results 2: processes of institution-building in Leipzig's media sector

New institutional structures which have developed in Leipzig support the growth of the media sector (Bathelt, 2002; Bathelt and Jentsch, 2002). Special training programs and institutions of higher education shape the local labor market and create specialized skills in media professions. Graduates from such institutions establish a basis for further firm formation which is, in turn, supported by start-up consulting and financial funds. Start-up decisions are also assisted by incubator organizations which provide space, services and organizational support for new firms in the media industry. Overall, a dense network of institutional support is developing within the region which provides a distinct institutional thickness (Amin and Thrift, 1995). This will likely serve to strengthen firm formation in the future.

Labor market, higher education and training programs

During the 1990s, a variety of new education and training programs which contribute to a specialized local labor market have been established by the university and other institutions of higher education in Leipzig (Sagurna, 1999, 2000). The graduates from these institutions create a specialized knowledge base in a variety of media-related fields. Students are able to gather practical experience during their studies and conduct research projects with local media firms, especially in the area of TV/film production and multimedia and internet applications. Many students also have part-time jobs or work during semester breaks in the media sector to finance their studies. The local media firms benefit from this exchange since it provides them with access to external knowledge and new conceptual ideas.

Start-up consulting

In order to create incentives for entrepreneurial activities, political and economic actors of the Leipzig region set up a combined start-up policy in the late 1990s. The Leipziger Sparkasse (a local bank), the City and County of Leipzig and an industry association established a joint office for start-up consulting and finance, the "ugb" (Unternehmensgründerbüro, Start-up Bureau). Since the ugb began its activities in 1998, it has supported more than 300 start-ups in the Leipzig area. According to the director of the ugb, about 350 jobs were created in 2000 through new ventures which had been supported by the ugb and received bank loans from Leipziger Sparkasse. Among these were also media firms in the areas of arts, graphics/design and marketing/communications.

Incubator centers

In 2000, the Media City Leipzig opened up as an incubator and technology center which provides office, workshop and studio space for about 70 TV and film-related firms and 500 employees (Schubert, 1999; Media City Leipzig, 2000). It is located adjacent to the MDR. Several interviewees pointed out that the MDR had put pressure on some subcontractors and suppliers to relocate into the new facilities of the Media City. In some cases, this seemed to have been a prerequisite for further contracts from the MDR. Some firms criticized this because they were not very pleased with this type of business conduct, as the rents for office space in the Media City were higher than those in other areas of Leipzig.

Another incubator facility for young firms and new services is the Business & Innovation Centre Leipzig which also hosts the ugb mentioned above. The Business & Innovation Centre was opened in 1999 and rents out office and laboratory space to young firms at low cost. In 2001, this facility had a total of 36 tenants, primarily in the areas of communication, marketing, design and new electronic services (Business & Innovation Centre Leipzig, 2001). The policy of this incubator facility is to support new firms for a five-year time period after which they should spin out to nearby locations.

In addition, other business centers exist which also support the local media sector (Schubert, 1999, 2000). The Medienhof Leipzig–Stötteritz, for instance, hosts more than 20 firms, primarily in traditional media branches.

Empirical results 3: regional market orientation and MDR's role as a major hub

The MDR and its activities have become the most important driving force behind the growth of Leipzig's media sector. This broadcasting network is the largest local customer of the film and TV industry and has attracted many service providers and suppliers (Sagurna, 1999, 2000). Throughout the 1990s, the local activities of the MDR continuously expanded. As one executive pointed out, the strategy employed was to spin out functions into separate subsidiaries and subcontract other functions to local suppliers. The goal behind this strategy was to establish a local supply and support sector for TV/film production and to cut costs. In addition, some suppliers and service providers from other regions were required to establish branches in Leipzig to keep their contracts. As a result of this policy, the MDR is no longer forced to rely on contractors located in different regions. Due to the increasing agglomeration of TV- and film-related businesses, about 70 percent of the MDR's production contracts are with suppliers and service providers from nearby locations (Reiter, 2000). This has, in turn, stimulated further start-up and relocation activities.

Thus far, the genesis of Leipzig's media industry appears to be a success story of industrial clustering. There are, however, severe limitations to this growth process. Many media firms have limited economic success which is indicated by low overall sales and growth rates in sales. More than a quarter of the respondents (15 of 56 firms) had annual sales of less than 150,000 euros (Table 9.3; see also Bentele *et al.*, 2000). This indicates that a substantial proportion of Leipzig's media firms do not earn a reasonable profit which, in turn, could stimulate further investments and future growth.

A reason for this can be found in the strong orientation towards local customers and the focus on small, often local market segments. A large share of the media firms surveyed almost exclusively sell their products and services to the regional market. They have hardly any external linkages to other market regions. Almost 60 percent of the media firms interviewed (36 of 62 respondents) sell more than two-thirds of their products and services within the Leipzig region (Table 9.4; see also Hinz, 1998). This seems to be a similar trend to that identified by Brail and Gertler (1999) in the case of Toronto's multimedia cluster. Closer analysis, however, reveals important differences between the two cases, as the Toronto economy hosts a number of fast-growing customer industries with national and international market reach.

In the case of Leipzig, the orientation towards the regional market is problematic because there is only limited regional growth potential: (i) the MDR completed its investment activities in the region and will not continue to grow at the same rate as in the past – due to financial irregularities in 2000, the MDR even reintegrated some tasks in-house; (ii) further, Leipzig's new convention center has problems competing against powerful competitors in West German cities, especially Frankfurt and Hanover – local contracts from the new convention center for firms in the areas of PR/marketing, exhibition and related TV/film services are thus lower than expected; (iii) finally, there is a lack of demand from the regional manufacturing sector which has not recovered from its previous blow associated with German Reunification.

Table 9.3 Media firms in Leipzig by sales, 2000/2001

Sales category	*Survey firms*	
	Number	*Share (%)*
Less than 50,000 euros	1	1.8
50,000–150,000 euros	14	25.0
150,000–400,000 euros	12	21.4
400,000–1,250,000 euros	22	39.3
1,250,000 euros and higher	7	12.5
Total	56	100.0

Source: Survey results

Table 9.4 Media firms in Leipzig by regional sales share, 2000/2001

Regional sales share	*Survey firms*	
	Number	*Share (%)*
Less than 33.3%	18	29.0
33.3–66.7%	8	12.9
66.7–100.0%	36	58.1
Total	62	100.0

Source: Survey results

Overall, this corresponds with the view that a missing link to external markets severely limits regional growth prospects (Scott, 1998; Maillat, 1998; Bathelt *et al.*, 2002). The strong orientation towards regional markets bears the risk of developing social relations which are characterized by over-embeddedness and stagnation (Uzzi, 1997). It seems that many of Leipzig's newly established media firms do not have the capability to develop pipelines to external markets.

Conclusions

The genesis of Leipzig's media sector appears to be a successful example of economic transformation. It has been shown that the growth of this sector is not a mere continuation of Leipzig's pre-war tradition in the book-publishing industry (Bathelt and Boggs, 2003). The former core sector was ruptured following the Second World War and, at the time of Reunification, was not prepared for market-led competition. However, regional re-bundling took place around the activities of the MDR, which served as a catalyst for the development of a new-media industry cluster. This cluster in the areas of TV/radio/film production, graphics/design, new digital media and related areas does not draw upon previous industrial structures and institutional settings.

In this chapter, I have analyzed the start-up and location processes associated with the growth of this cluster and the processes of institution-building. This has been done using a multidimensional cluster approach which distinguishes clusters along their horizontal, vertical, institutional, external and power dimensions. As clusters do not exist in their final form from the very beginning but gradually develop over time, an evolutionary perspective has to be used to understand the underlying social and economic processes. In the case of Leipzig, it has been shown in which ways new media firms have been started up, how they have shaped the local labor market, and how this process has contributed to and benefited from the formation of specialized media-related institutions. The multidimensional cluster framework used allows us to evaluate different facets of this growth process.

Until today, Leipzig's media sector cannot be viewed as a fully developed industry cluster, as several dimensions are still in an infant stage. The start-up processes observed have contributed to the growth in the vertical cluster dimension. Meanwhile, there is a variety of specialized suppliers and service providers. Regional producers are no longer forced to acquire the majority of their supplies from contractors located in other regions. In contrast, the horizontal cluster dimension is still underdeveloped. There are only few large customers within the region and there is a lack of TV and film production companies.

The most important trigger for the development of this cluster was the political decision to locate the MDR head office and production facilities to Leipzig. The MDR has stimulated the development of the local media industry in many different ways: (i) as an important institution which defines rules and formats in the TV and film business; (ii) as a customer which attracts other media firms and stimulates start-up processes; and (iii) through its policy to favor local subcontractors and suppliers. Parallel to this development, local policy initiatives were designed to provide start-up consulting, refine training programs and create new incubator facilities for entrepreneurial activities. These have contributed to a growing institutional dimension and could help produce more buzz in the future. It would be wrong, however, to reduce the emergence of Leipzig's media industry cluster solely to political influence and the role of the MDR. Notably, not all of the media branches are closely related to the TV and film business.

At this point, however, the impact of Leipzig's media sector on the local labor market is still limited. Most firms are quite small with only few employees and modest sales. They do not earn a reasonable profit which, in turn, could stimulate further investments. A reason for this can be found in the underdeveloped external dimension of the cluster. Firms have not been able to establish trans-local pipelines which provide access to new markets. If this does not change this could be a severe burden for the future development of Leipzig's media sector. A wave of bankruptcies could occur in the upcoming years (Thomas, 2001) if markets do not grow substantially. There are, however, also some positive prospects. Porsche's and BMW's recent decisions to establish new automobile-production facilities in the region (Wüpper, 2002) help to strengthen the local economy and stimulate additional growth from which some segments of the media sector could benefit.

Acknowledgments

This chapter is based on previous work by the author and collaborations with various colleagues (Bathelt, 2001, 2002, 2003; Bathelt and Jentsch, 2002; Bathelt and Taylor, 2002; Bathelt *et al.*, 2002; Bathelt and Boggs, 2003). I would like to thank the editors of this volume, as well as one anonymous reviewer and Pat McCurry, for their valuable suggestions which have helped to make this chapter more compact.

Bibliography

Allen, J. (1997) "Economies of power and space," in Lee, R. and Wills, J. (eds) *Geographies of Economies*, pp. 59–70, London: Arnold.

Amin, A. and Thrift, N. (1995) "Living in the global," in Amin, A. and Thrift, N. (eds) *Globalization, Institutions, and Regional Development in Europe*, pp. 1–22, Oxford and New York: Oxford University Press.

Arthur, W. B. (1988) "Competing technologies: an overview," in Dosi, G., Freeman, C., Nelson, R. R., Silverberg, G. and Soete, L. L. G. (eds) *Technical Change and Economic Theory*, pp. 590–607, London and New York: Pinter.

Baier, H. (ed.) (1992) *Medienstadt Leipzig: Tradition and Perspektiven (Leipzig's Media Industry: Traditions and Perspectives)*, Berlin: Vistas.

Bathelt, H. (2001) *The Rise of a New Cultural Products Industry Cluster in Germany: The Case of the Leipzig Media Industry*, IWSG Working Papers 06–2001, Frankfurt am Main; available online: <http://www.rz.uni-frankfurt.de/FB/fb18/wigeo/iwsg.html> (accessed 11 October 2003).

—— (2002) "The re-emergence of a media industry cluster in Leipzig," *European Planning Studies*, 10: 583–611.

—— (2003) "Success in the local environment: local buzz, global pipelines and the importance of clusters," *think on*, no. 2: 28–33; available online: <http://www.altana.com/root/index.php?lang=en&page_id=893> (accessed 11 October 2003).

Bathelt, H. and Boggs, J. S. (2003) "Towards a re-conceptualization of regional development paths: is Leipzig's media cluster a continuation of or a rupture with the past?," *Economic Geography*, 79: 265–93.

Bathelt, H. and Jentsch, C. (2002) "Die Entstehung eines Medienclusters in Leipzig: Neue Netzwerke und alte Strukturen" [The genesis of a new media industry cluster in Leipzig: new networks and old structures], in Gräf, P. and Rauh, J. (eds) *Networks and Flows: Telekommunikation zwischen Raumstruktur, Verflechtung und Informationsgesellschaft*, pp. 31–74, Hamburg and Münster: Lit.

Bathelt, H., Malmberg, A. and Maskell, P. (2002) "Clusters and knowledge: local buzz, global pipelines and the process of knowledge creation," DRUID Working Paper 2002–12, Copenhagen; available online: <http://www.druid.dk/wp/wp.html> (accessed 11 October 2003); and also in *Progress in Human Geography*, 28: 31–56, 2004.

Bathelt, H. and Taylor, M. (2002) "Clusters, power and place: inequality and local growth in time–space," *Geografiska Annaler*, 84 B: 93–109.

Bentele, G., Liebert, T. and Polifke, M. (2000) *Medienstandort Leipzig III: Eine Studie zur Leipziger Medienwirtschaft 2000 [Leipzig as a Location of the Media Industry III: A Study of Leipzig's Media Economy 2000]*, Leipzig: Medienstadt.

Bentele, G., Polifke, M. and Liebert, T. (1998) *Medienstandort Leipzig II: Eine Studie zur Leipziger Medienwirtschaft 1998 [Leipzig as a Location of the Media Industry II: A Study of Leipzig's Media Economy 1998]*, Leipzig: Medienstadt.

Boggs, J. S. (2001) "Path dependency and agglomeration in the German book publishing industry," paper presented at the Annual Meeting of the Association of American Geographers, New York.

Boyer, R. (1988) "Technical change and the theory of 'régulation'," in Dosi, G., Freeman, C., Nelson, R. R., Silverberg, G. and Soete, L. L. G. (eds) *Technical Change and Economic Theory*, pp. 67–94, London and New York: Pinter.

—— (1990) *The Regulation School: A Critical Introduction*, New York: Columbia University Press.

Brail, S. G. and Gertler, M. S. (1999) "The digital regional economy: emergence and evolution of Toronto's multimedia cluster," in Braczyk, H.-J., Fuchs, G. and Wolf, H.-G. (eds) *Multimedia and Regional Economic Restructuring*, pp. 97–130, London and New York: Routledge.

Bramanti, A. and Ratti, R. (1997) "The multi-faced dimensions of local development," in Ratti, R., Bramanti, A. and Gordon, R. (eds) *The Dynamics of Innovative Regions: The GREMI Approach*, pp. 3–44, Aldershot and Brookfield: Ashgate.

Business & Innovation Centre Leipzig (2001) "Präsentation des Business & Innovation Centre Leipzig" [Presentation of the Business & Innovation Centre Leipzig], unpublished paper presentation, Leipzig.

Clark, G. L. and Tracey, P. (2004) *Global Competitiveness and Innovation: An Agent-Centred Perspective*, Houndsmill, New York: Palgrave Macmillan.

Clegg, S. (1989) *Frameworks of Power*, London: Sage.

Cohen, W. M. and Levinthal, D. A. (1990) "Absorptive capacity: a new perspective on learning and innovation," *Administrative Science Quarterly*, 35: 128–52.

Cooke, P. and Morgan, K. (1998) *The Associational Economy*, Oxford: Oxford University Press.

Crevoisier, O. and Maillat, D. (1991) "Milieu, industrial organization and territorial production system: towards a new theory of spatial development," in Camagni, R. (ed.) *Innovation Networks: Spatial Perspectives*, pp. 13–34, London and New York: Belhaven Press.

Crewe, L. (1996) "Material culture: embedded firms, organizational networks and the local economic development in a fashion quarter," *Regional Studies*, 30: 257–72.

Denzer, V. and Grundmann, L. (1999) "Das Graphische Viertel – ein citynahes Mischgebiet der Stadt Leipzig im Transformationsprozeß: Vom Druckgewerbe- zum Bürostandort" [The transformation of Leipzig's Graphical Quarter: from printing and publishing to modern office functions], *Europa Regional*, 7 (3): 37–50.

Deutsches Institut für Wirtschaftsforschung (1999) *Perspektiven der Medienwirtschaft (Perspectives of the Media Economy)*, Berlin.

Egan, E. A. and Saxenian, A. (1999) "Becoming digital: sources of localization in the Bay area multimedia cluster," in Braczyk, H.-J., Fuchs, G. and Wolf, H.-G. (eds) *Multimedia and Regional Economic Restructuring*, pp. 11–29, London and New York: Routledge.

Fujita, M., Krugman, P. and Venables, A. J. (1999) *The Spatial Economy. Cities, Regions and International Trade*, Cambridge, MA: MIT Press.

Gertler, M. S. (1993) "Implementing advanced manufacturing technologies in mature industrial regions: towards a social model of technology production," *Regional Studies*, 27: 665–80.

—— (1995) "'Being there": proximity, organization, and culture in the development and adoption of advanced manufacturing technologies," *Economic Geography*, 71: 1–26.

—— (1997) "The invention of regional culture," in Lee, R. and Wills, J. (eds) *Geographies of Economies*, pp. 47–58, London and New York: Arnold.

Gordon, I. R. and McCann, P. (2000) "Industrial clusters: complexes, agglomeration and/or social networks," *Urban Studies*, 37: 513–32.

Gormsen, N. (1996) *Leipzig – Stadt, Handel, Messe: Die städtebauliche Entwicklung der Stadt Leipzig als Handels- und Messestadt (Leipzig's Development as a Trade and Exhibition Center)*, Leipzig: Institut für Länderkunde.

Grabher, G. (1993) "Rediscovering the social in the economics of interfirm relations," in Grabher, G. (ed.) *The Embedded Firm. On the Socioeconomics of Industrial Networks*, pp. 1–31, London and New York: Routledge.

—— (2001) "Ecologies of creativity: the village, the group, and the heterarchic organisation of the British advertising industry," *Environment and Planning A*, 33: 351–74.

—— (2002) "Cool projects, boring institutions: temporary collaboration in social context," *Regional Studies*, 36: 205–14.

Gräf, P. (2001) "Das Buchverlagswesen und seine Standorte" [The German book publishing industry and its locations], in Institut für Länderkunde (eds) *Nationalatlas Bundesrepublik Deutschland: Band 9. Verkehr und Kommunikation*, pp. 116–17, Heidelberg and Berlin: Spektrum.

Gräf, P. and Matuszis, T. (2001) "Medienstandorte: Schwerpunkte und Entwicklungen" [German media centers and their development], in Institut für Länderkunde (eds) *Nationalatlas Bundesrepublik Deutschland: Band 9. Verkehr und Kommunikation*, pp. 114–15, Heidelberg and Berlin: Spektrum.

Gräf, P., Hallati, H. and Seiwert, P. (2001) "Öffentlich-rechtliche und private Rundfunk- und Fernsehanbieter" [Public and private television and broadcasting services in Germany], in Institut für Länderkunde (eds) *Nationalatlas Bundesrepublik Deutschland: Band 9. Verkehr und Kommunikation*, pp. 118–21, Heidelberg and Berlin: Spektrum.

Granovetter, M. (1973) "The strength of weak ties," *American Journal of Sociology*, 78: 1360–80.

—— (1985) "Economic action and economic structure: the problem of embeddedness," *American Journal of Sociology*, 91: 481–510.

Grundmann, L. (1996) "Die Leipziger City im Wandel – zwischen der Tradition als Messe- und Handelsplatz und aktueller Innenstadtentwicklung" [Changes in Leipzig's city center: traditional functions and new developments], in Grundmann, L., Tzschaschel, S. and Wollkopf, M. (eds) *Leipzig: Ein geographischer Führer durch Stadt und Umland*, pp. 30–55, Leipzig: Thom.

Hayter, R. (1997) *The Dynamics of Industrial Location: The Factory, the Firm and the Production System*, Chichester and New York: Wiley.

Hinz, T. (1998) *Betriebsgründungen in Ostdeutschland [Firm Start-ups in East Germany]*, Berlin: Edition Sigma – Bohn.

Hinz, T. and Ziegler, R. (2000) "Ostdeutsche Gründerzentren revisited: Eine Bilanz 10 Jahre nach dem Fall der Mauer" [East German start-up centers revisited: a review 10 years after the fall of the Berlin Wall], in Esser, H. (ed.) *Der Wandel nach der Wende: Gesellschaft, Wirtschaft, Politik in Ostdeutschland*, pp. 237–50, Wiesbaden: Westdeutscher Verlag.

Hodgson, G. M. (1988) *Economics and Institutions: A Manifesto for a Modern Institutional Economics*, Cambridge: Polity Press.

Jöns, H. (2001) "Foreign banks are branching out: changing geographies of Hungarian banking, 1987–1999," in Meusburger, P. and Jöns, H. (eds) *Transformations in Hungary. Essays in Economy and Society*, pp. 65–124, Heidelberg and New York: Physica.

Kern, H. (1996) "Vertrauensverlust und blindes Vertrauen: Integrationsprobleme im ökonomischen Handeln" [Loss of trust and blind confidence in economic action], *SOFI-Mitteilungen*, 24: 7–14.

Kowalke, H. (1994) "Wirtschaftsraum Sachsen" [Economic structure of Saxony], *Geographische Rundschau*, 46: 484–90.

Krätke, S. (2002) *Medienstadt: Urbane Cluster und globale Zentren der Kulturproduktion [Media Cities: Urban Centers of Cultural Production]*, Opladen: Leske + Budrich.

Krätke, S. and Scheuplein, C. (2001) *Produktionscluster in Ostdeutschland: Methoden der*

Identifizierung und Analyse [Production Clusters in East Germany: Methods of Identification and Analysis], Hamburg: VSA.

Krugman, P. (1991) *Geography and Trade*, Leuven: Leuven University Press and Cambridge, MA/London: MIT Press.

—— (2000) "Where in the world is the 'new economic geography'?," in Clark, G. L., Feldman, M. P. and Gertler, M. S. (eds) *The Oxford Handbook of Economic Geography*, pp. 49–60, Oxford: Oxford University Press.

Lash, S. and Urry, J. (1994) *Economies of Signs and Spaces*, London: Sage.

Latour, B. (1986) "The powers of association," in Law, J. (ed.) *Power, Action and Belief: A New Sociology of Knowledge?*, pp. 264–80, London: Routledge & Kegan Paul.

Lawson, C. (1999) "Towards a competence theory of the region," *Cambridge Journal of Economics*, 23: 151–66.

Lawson, C. and Lorenz, E. (1999) "Collective learning, tacit knowledge and regional innovative capacity," *Regional Studies*, 33: 305–17.

Leyshon, A., Matless, D. and Revill, G. (1998) *The Place of Music*, New York: Guilford.

Lipietz, A. (1987) *Mirages and Miracles: The Crises of Global Fordism*, London: Verso.

Lorenz, E. (1999) "Trust, contract and economic cooperation," *Cambridge Journal of Economics*, 23: 301–15.

Lundvall, B.-Å. (1988) "Innovation as an interactive process: from user–producer interaction to the national system of innovation," in Dosi, G., Freeman, C., Nelson, R. R., Silverberg, G. and Soete, L. L. G. (eds) *Technical Change and Economic Theory*, pp. 349–69, London: Pinter.

Maillat, D. (1998) "Vom 'Industrial District' zum innovativen Milieu: Ein Beitrag zur Analyse der lokalen Produktionssysteme" (From industrial districts to innovative milieus: towards an analysis of territorial production systems), *Geographische Zeitschrift*, 86: 1–15.

Malecki, E. J. (2000) "Knowledge and regional competitiveness," *Erdkunde*, 54: 334–51.

Malmberg, A. and Maskell, P. (2002) "The elusive concept of localization economies: towards a knowledge-based theory of spatial clustering," *Environment and Planning A*, 34: 429–49.

Marshall, A. (1920) *Principles of Economics*, 8th edn, Philadelphia: Porcupine Press.

—— (1927) *Industry and Trade. A Study of Industrial Technique and Business Organization; and Their Influences on the Conditions of Various Classes and Nations*, 3rd edn, London: Macmillan.

Maskell, P. (2001) "Towards a knowledge-based theory of the geographical cluster," *Industrial and Corporate Change*, 10: 921–43.

Maskell, P. and Malmberg, A. (1999a) "The competitiveness of firms and regions: 'ubiquitification' and the importance of localized learning," *European Urban and Regional Studies*, 6: 9–25.

—— (1999b) "Localised learning and industrial competitiveness," *Cambridge Journal of Economics*, 23: 167–85.

Media City Leipzig (2000) *Media City Leipzig: Zentrum für elektronische Medien [Media City Leipzig: Center of New Electronic Media Branches]*, Leipzig.

Murdoch, J. (1995) "Actor–networks and the evolution of economic forms: combining description and explanation in theories of regulation, flexible specialization, and networks," *Environment and Planning A*, 27: 731–57.

North, D. C. (1990) *Institutions, Institutional Change and Economic Performance*, Cambridge: Cambridge University Press.

Oelke, E. (1997) *Sachsen-Anhalt [Saxony-Anhalt]*, Gotha: Perthes.

Oinas, P. (1997) "On the socio-spatial embeddedness of business firms," *Erdkunde*, 51: 23–32.

Owen-Smith, J. and Powell, W. W. (2002) "Knowledge networks in the Boston biotechnology community," paper presented at the conference on "Science as an Institution and the Institutions of Science," Siena.

Porter, M. E. (1990) *The Competitive Advantage of Nations*, New York: Free Press.

—— (1998) "Clusters and the new economics of competition," *Harvard Business Review*, 76 (November-December): 77–90.

—— (2000) "Locations, clusters, and company strategy," in Clark, G. L., Feldman, M. P. and Gertler, M. S. (eds) *The Oxford Handbook of Economic Geography*, pp. 253–74, Oxford: Oxford University Press.

Power, D. (2002) "'Cultural industries' in Sweden: an assessment of their place in the Swedish economy," *Economic Geography*, 78: 103–27.

Pratt, A. (1997) "Cultural industries: guest editorial," *Environment and Planning A*, 29: 1911–17.

Reiter, U. (2000) "Der MDR in der Medienstadt Leipzig" [The role of the MDR in Leipzig], in Grunau, H., Kleinwächter, W. and Stiehler, H.-J. (eds) *Medienstadt Leipzig: Vom Anspruch zur Wirklichkeit*, pp. 37–40, Leipzig: Monade.

Sagurna, M. (1999) "Medienstandort Sachsen – Bestandsaufnahme und Perspektiven" [Saxony's media industry – present state and future perspectives], in Altendorfer, O. and Mayer, K.-U. (eds) *Sächsisches Medienjahrbuch 1998/1999*, pp. 12–19, Leipzig: Verlag für Medien & Kommunikation.

—— (2000) "Der Medienstandort Leipzig im Freistaat Sachsen" [Leipzig's role as a media location in Saxony], in Grunau, H., Kleinwächter, W. and Stiehler, H.-J. (eds) *Medienstadt Leipzig: Vom Anspruch zur Wirklichkeit*, pp. 22–30, Leipzig: Monade.

Schmidt, H. (1994) "Leipzig zwischen Tradition und Neuorientierung" [The city of Leipzig: traditional structures and re-orientation], *Geographische Rundschau*, 46: 500–7.

Schubert, D. (1999) "Media City – Leipzig: Bestandsaufnahme und Ausblick" [Media City Leipzig: present state and future perspectives], in Altendorfer, O. and Mayer, K.-U. (eds) *Sächsisches Medienjahrbuch 1998/1999*, pp. 20–3, Leipzig: Verlag für Medien & Kommunikation.

—— (2000) "Die Stadt Leipzig und die Medien als Wirtschaftsfaktor" [The city of Leipzig and the economic importance of media], in Grunau, H., Kleinwächter, W. and Stiehler, H.-J. (eds) *Medienstadt Leipzig: Vom Anspruch zur Wirklichkeit*, pp. 33–6, Leipzig: Monade.

Schulz, G. (1989) *Buchhandels-Ploetz [Ploetz Book Trade Directory]*, 4th edn, Freiburg: Ploetz.

Scott, A. J. (1996) "The craft, fashion, and cultural-products industries of Los Angeles: competitive dynamics and policy dilemmas in a multisectoral image-producing complex," *Annals of the Association of American Geographers*, 86: 306–23.

—— (1988) *New Industrial Spaces. Flexible Production Organization and Regional Development in North America and Western Europe*, London: Pion.

—— (1998) *Regions and the World Economy: The Coming Shape of Global Production, Competition, and Political Order*, Oxford and New York: Oxford University Press.

—— (2000) *The Cultural Economy of Cities: Essays on the Geography of Image-Producing Industries*, London, Thousand Oaks, New Delhi: Sage.

Scott, J. (2001) *Power*, Cambridge and Oxford: Polity Press.

Sofer, M. and Schnell, I. (2002) "Over- and under-embeddedness: failures in developing mixed embeddedness among Israeli Arab entrepreneurs," in Taylor, M. and Leonard, S. (eds) *Embedded Enterprise and Social Capital: International Perspectives*, pp. 207–24. Aldershot: Ashgate.

Statistisches Landesamt des Freistaates Sachsen (2001) *Umsätze und ihre Besteuerung im Freistaat Sachsen: Ergebnisse der Umsatzsteuerstatistik 1999 [Tax Revenue Statistics 1999]*, project-specific analysis, Kamenz.

Storper, M. (1995) "The resurgence of regional economics, ten years later," *European Urban and Regional Studies*, 2: 191–221.

—— (1997) *The Regional World. Territorial Development in a Global Economy*, New York and London: Guilford.

Storper, M. and Christopherson, S. (1987) "Flexible specialization and regional industrial agglomerations: the case of the U.S. motion-picture industry," *Annals of the Association of American Geographers*, 77: 260–82.

Storper, M. and Venables, A. J. (2002) "Buzz: the economic force of the city," paper presented at the DRUID summer conference on "Industrial Dynamics of the New and Old Economy – Who is Embracing Whom?," Copenhagen, Elsinore.

Storper, M. and Walker, R. (1989) *The Capitalist Imperative. Territory, Technology, and Industrial Growth*, New York and Oxford: Basil Blackwell.

Sydow, J. and Staber, U. (2002) "The institutional embeddedness of project networks: the case of content production in German television," *Regional Studies*, 36: 215–27.

Taylor, M. (1995) "The business enterprise, power and patterns of geographical industrialisation," in Conti, S., Malecki, E. J. and Oinas, P. (eds) *The Industrial Enterprise and its Environment: Spatial Perspectives*, pp. 99–122, Aldershot: Ashgate.

—— (2000) "Enterprise, power and embeddedness: an empirical exploration," in Vatne, E. and Taylor, M. (eds) *The Networked Firm in a Global World: Small Firms in New Environments*, pp. 199–233, Aldershot, Burlington: Ashgate.

Thomas, M. (2001) *Ein Blick zurück und voraus: Ostdeutsche Neue Selbständige – aufgeschobenes Scheitern oder Potenziale zur Erneuerung? [Looking Back and Forth: New Self-Employment in East Germany – Postponed Failure or Potential for Renewal?]*; available online: <http://www.biss-online.de/index_html.htm> (accessed October 11, 2003).

Tracey, P. and Clark, G. L. (2003) "Alliances, networks and competitive strategy: rethinking clusters of innovation," *Growth and Change*, 34: 1–16.

Uzzi, B. (1997) "Social structure and competition in interfirm networks: the paradox of embeddedness," *Administrative Science Quarterly*, 42: 35–67.

Wittmann, R. (1999) *Geschichte des deutschen Buchhandels [History of the German Book Trade]*, 2nd edn, München: Beck.

Wüpper, T. (2002) "Porsche und BMW sollen erst der Anfang sein" [Porsche and BMW are just the beginning], *Frankfurter Rundschau*, August 17, p. 9.

Yeung, H. W.-c. (1994) "Critical reviews of geographical perspectives on business organizations and the organization of production: towards a network approach," *Progress in Human Geography*, 18: 460–90.

10 Manufacturing culture in Birmingham's Jewelry Quarter

Jane Pollard

Introduction

A number of social-science disciplines have undergone variants of "a cultural turn" since the 1980s, both in terms of opening dialogues with post-structural, feminist, literary and ecological literatures and in terms of conceptualizing the social and cultural construction of economic practices (Thrift and Olds, 1996; Crang, 1997; Peet, 1997). Part of the impetus for this shift is the growing significance and visibility of cultural outputs and employment in contemporary capitalist economies. Although much recent research on so-called "cultural industries" has considered developments in services, media, fashion, entertainment and music, this chapter focuses on recent developments in jewelry production in the U.K.'s second city, Birmingham.

Although dwarfed by the U.S. jewelry industry (which had sales of $53 billion in 2002), the U.K. jewelry industry vies with Italy as Europe's largest with sales of $5.2 billion (£3.3 billion) in 2002 (http://www.signetgroupplc.com/signetplc/about/operatingrev/marketplace/, accessed 6 November 2003). As one of the oldest manufacturing industries in the Birmingham and West Midlands region, the jewelry industry is an interesting site from which to explore the significance now attached to cultural production and consumption in urban-regeneration schemes. Specifically, the chapter examines two attempts to rejuvenate the industry, first, by promoting it as a modern, design-led industry and, second, by exploiting the tourist potential of Birmingham's historic Jewelry Quarter.

This case study is developed in order to make three points about rejuvenating the industry. First, it is important not only to understand policy agendas but also to examine the experiences of those firm owners and workers who live through such policy initiatives (see Kong, 2000). Second, for all the emphasis on the symbolic, aesthetic, semiotic and positional functions of cultural products, it is important to examine the material and geographically rooted bases for the production of such commodities. Recent developments in the Jewelry Quarter are useful vehicles for revealing some of the material conditions required to produce creativity and provide a useful counterpoint to simplistic accounts that either eulogize or dismiss creative industries (Crewe *et al.*, 2003).

The third point is an observation concerning one of the hallmarks of much contemporary social-science research, namely an insistence on the social and cultural construction of the economic (Barnes, 1996; Thrift and Olds, 1996; Lee and Wills, 1997; Sayer, 2001). Economic geographers, sociologists and institutional and evolutionary economists have generated a compelling critique of the under-socialized view of human agency that typifies much mainstream economic theory. Policy makers in many arenas have been pushed to explore and understand the social, cultural and material assets of production in order to imitate "successful" production systems in different places/sectors. Birmingham's Jewelry Quarter, however, exemplifies some of the contradictory, contested qualities of networks of power, status and identity that are bound up with ingrained understandings and cultures of work. These networks have both attracted and excluded new entrants into the industry and, in the face of unrelenting competitive pressures, have sometimes hampered change, adaptation and trust building. Moving beyond an acknowledgement that all industries are "cultural," this chapter argues that reinvigorating Birmingham's Jewelry Quarter will involve more than improving its physical surroundings and nurturing its considerable pool of design talent. Encouraging jewelers to change their ways of working, to embrace design and to move into higher value-added markets will also involve understanding existing values, and, as Schoenberger argues, (1997) the processes of valuation that legitimate the status quo, even when that status quo is one of decline.

The chapter begins with some context – providing a brief overview of Birmingham's recent redevelopment initiatives – before considering the development of the Jewelry Quarter and two specific initiatives aimed at rejuvenating the industry and the Quarter. The chapter then examines some of the tensions and contradictions in these initiatives, drawing on a range of sources including published and unpublished statistics and over thirty interviews with jewelry manufacturers, designer-makers and key informants in a range of public, private and voluntary institutions in Birmingham. Finally, the chapter considers some of the silences of these policies and highlights a series of problems that challenge prevailing manufacturing cultures and ways of doing business in the Quarter.

The economics of a cultural turn in Birmingham

Cities, as places of social interaction, of dense divisions of labor, of knowledge flows, reflexivity and creativity (Storper, 1997; Scott, 2001; Florida, 2002), have long been key sites of cultural production and consumption. More recently, however, cities are also increasingly important, economically, politically and symbolically, in regenerating areas affected by manufacturing decline (Bianchini, 1993; Gdaniec, 2000). "Culture," it seems, "is now seen as the magic substitute for all the lost factories and warehouses, and as a device that will create a new urban image, making the city more attractive to mobile capital and mobile professional workers" (Hall, 2000: 640). Culture, however defined, is now big business in the U.K. Arguments about supporting cultural endeavors in terms of their aesthetic or social importance are now accompanied by arguments about the economic significance of what are now being termed "creative industries." The "creative industries" are a statistical aggregate generated by the U.K.'s Department for Culture, Media and Sport in 1998 (DCMS 1998). The DCMS defined the "creative industries" as

> those industries which have their origin in individual creativity, skill and talent and which have a potential for wealth and job creation through the generation and exploitation of intellectual property. This includes advertising, architecture, the art and antiques market, crafts, design, designer fashion, film and video, interactive leisure software, music, the performing arts, publishing, software and computer services, television and radio.
>
> (http://www.culture.gov.uk/creative_industries/default.htm, accessed 10 November 2003)

This diverse assortment of industries has been growing rapidly in the U.K., although the vagaries of industrial classifications hinder precise measurement. Between 1997 and 2000 these industries' exports grew at about 13 percent per annum, while employment grew at five percent per annum, compared with one and a half percent for the whole economy (http://www.culture.gov.uk/creative_industries/default.htm_accessed, 23 April 2003). The DCMS estimate that the creative industries now constitute about four percent of U.K. GDP and generate £7.5 billion in exports (http://www.arts.org.uk/directory/regions/west_mid/ar99/p7.htm, accessed 23 April 2003).

Birmingham, a post-war center for employment in autos, engineering and metal goods, has been home to a series of service, leisure and retail projects intended to combat economic decline since the 1970s. Like other major U.K. city economies Birmingham experienced significant losses in manufacturing employment, declining population in central areas and a deteriorating physical fabric through the 1970s and 1980s (see Loftman and Nevin, 1996; Henry *et al.*, 2002; Pollard, 2004). The City Council have attracted lottery funds, European Regional Development funds and other public and private monies to redevelop the city's canals, pedestrianize parts of the city center and to build a series of flagship projects (Hubbard, 1996; Loftman and Nevin 1996; Ward, 2001; Webster, 2001; Henry *et al.*, 2002), including the National Exhibition Centre (NEC), the National Indoor Arena (NIA), the International Convention Centre (ICC), Birmingham Symphony Hall and Orchestra, Birmingham Royal Ballet/Repertory Theatre, Millennium Point, a series of retail and leisure complexes and a cultural and learning quarter ("Eastside").

These leisure, retail, educational and cultural developments dovetail not only with the City Council's desire to attract inward investment but also their ambition to increase the attractiveness of city-center living. The Council's "City Living Initiative" has promoted the development of over 2000 housing units between 1998 and 2003 and more are planned to help meet the council's target of an estimated 44,500 new homes required in Birmingham between 1991 and 2011 (EDAW, 1998). Thus far, these developments have produced accommodation for high-earning professionals and investors; an estimated 50 percent of the new flats built in various city-center regeneration schemes have been bought up by investors (Norwood, 2002).

How do Birmingham's established, and in many cases declining, manufacturing industries fit with this consumption led style of regeneration? If Birmingham's industrial past is to be repackaged and its cultural content highlighted, then jewelry appears to be a good candidate for such treatment. The industry has both volume and craft production, but in its craft variant relies on highly skilled workers and a strong design ethos. Before considering this issue in more detail, the following section describes the state of the trade in the Jewelry Quarter today.

Birmingham's place in the U.K. jewelry industry

Birmingham has been an internationally renowned center of jewelry design and manufacture since the eighteenth century (Roche, 1927; Wise, 1949; Mason, 1998), when goldsmiths and silversmiths, gemstone and bullion dealers and other jewelry-related traders began to agglomerate less than a mile to the northwest of the city center (see Figure 10.1) in Hockley. U.K. employment in jewelry production stands at just over 9,000 and Birmingham accounts for about 50 percent of this total; the other major centers of production are London and Sheffield (DTI, 2001; JQRP, 2002).

The jewelry industry is one of the most highly geographically concentrated industries in the U.K. (Devereaux *et al.*, 1999) and is characterized by large numbers of small firms, employing fewer than 10 people. One crude indicator of jewelry output is provided by hallmarking figures;[1] these reveal the peaks and troughs of a trade that is reliant on discretionary spending. For much of the twentieth century, Birmingham's Assay Office has hallmarked between 40–50 percent of all articles hallmarked in Britain (see Figure 10.2).

The Quarter today covers over 265 acres in Hockley and is home to over 500 jewelry-related firms (Pollard, 2004). Outside the Jewelry Quarter, the city's population of jewelers also includes a group of Asian jewelry retailers found, predominantly, in the Sparkhill area of the city. Estimates of the size of this community, the extent and nature of their production networks and their degree (or not) of linkage with the Jewelry Quarter are anecdotal and in need of further research.

Although the Jewelry Quarter remains a significant European center of jewelry design and production it is a shadow of its former self in terms of employment and number of firms. At its peak of employment in 1913, between 50,000 (Roche, 1927) and 70,000 workers (Smirke, 1913, cited in Mason 1998) were employed in jewelry production, second only to employment in the non-ferrous metal trades in Birmingham. Into the 1950s, there were over 1,500 firms and more than 30,000 people employed in the Quarter (Mason, 1998).

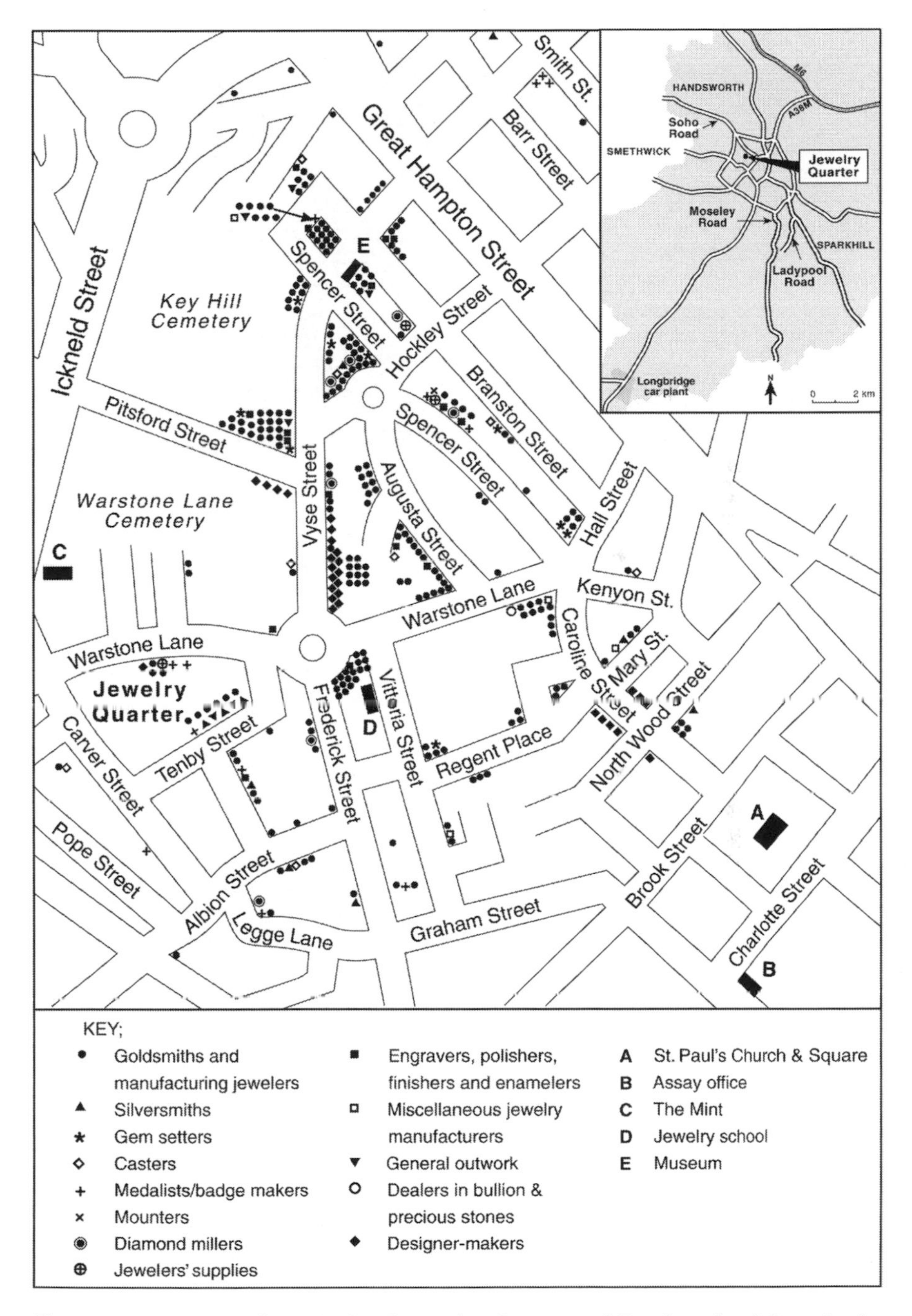

Figure 10.1 Location of Birmingham's Jewelry Quarter and Jewelry-related firms in the Quarter, 2001 (source: Business Link database, author's survey).

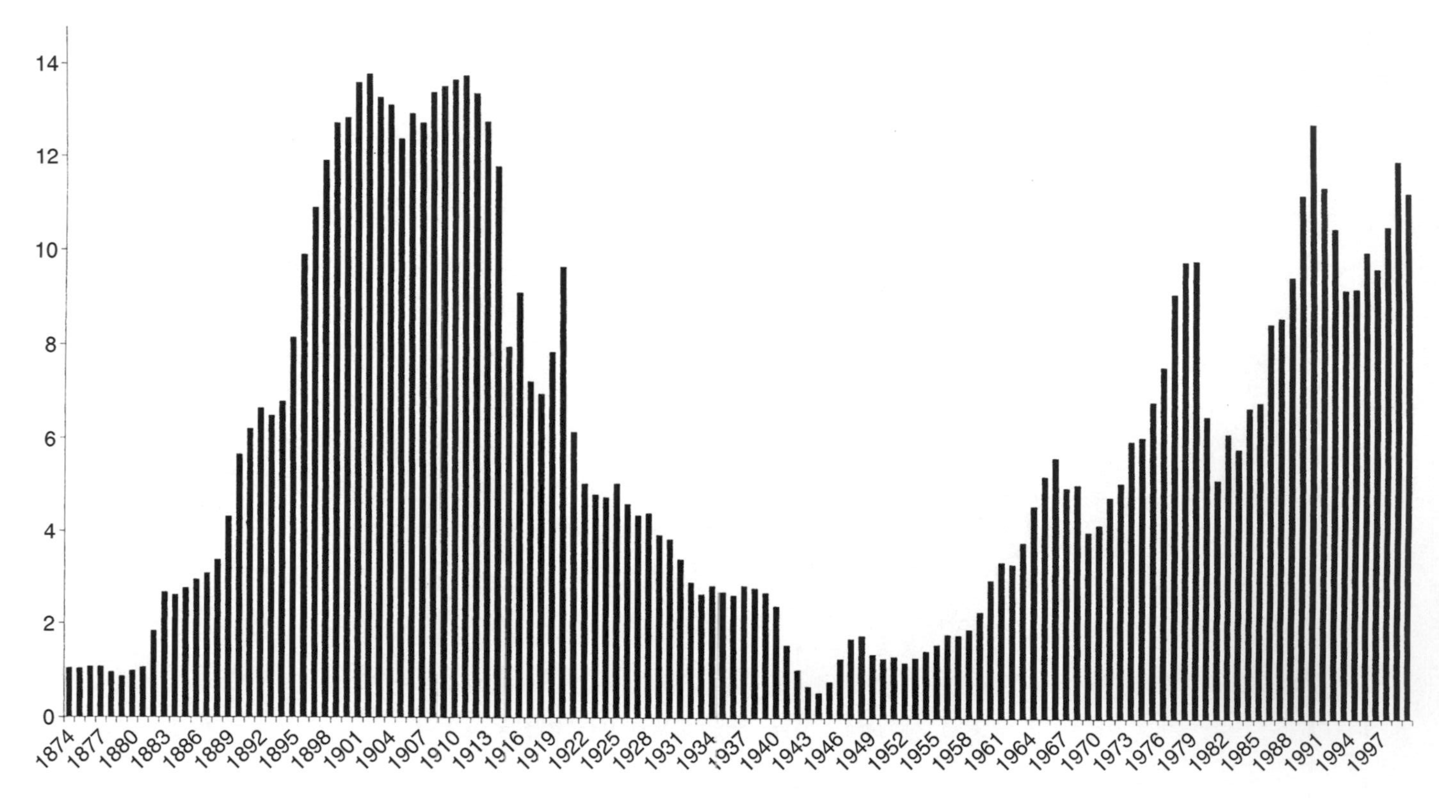

Figure 10.2 Millions of articles assayed in Birmingham, 1874–1999 (source: Assay office, unpublished statistics).

The Quarter grew rapidly through the nineteenth century, fuelled by discoveries of gold, the commercial adaptation of electroplating, the rising disposable income and demand for jewelry of the Victorian middle classes and legislation (the Lower Standards Act of 1854) facilitating the introduction of cheaper nine-, twelve- and fifteen-carat gold standards. The Jewelry Quarter also became the home to a range of jewelry-related institutions, notably an Assay Office (1773), the Birmingham Jewelers and Silversmiths Association (hereafter BJSA)(1887),[3] and a School of Jewelry (1890).

Although some Birmingham manufacturers have a tradition of designing and producing some exceptionally high-quality jewelry, many manufacturers embraced mass production in the post-war period. In this respect, Birmingham producers largely mirrored the U.K. industry in producing affordable nine-carat gold pieces for high-volume markets. Nevertheless, post-war conditions were initially difficult for jewelers; high purchase tax, the loss of labor to the growing, and usually better paid, automobile and engineering industries in the West Midlands, and the shortage of precious metals all hurt the industry. More significant has been the emergence of formidable overseas competition on two fronts through the 1980s and 1990s. In high value-added markets are competitors from Switzerland, Germany and most especially Italy; Italy exported 75 percent of its jewelry in 1998, accounting for almost 25 percent of world jewelry exports (by value) (DTI, 2001). At the lower end of the market, lower wage producers such as Thailand, Indonesia, China and India have made significant inroads into the U.K. nine-carat market and eroded U.K. manufacturers' market share (from 68 percent in 1980 to 56 percent in 1999) (DTI, 2001). Against this context of decline since the 1980s, a constituency of "designer-maker" firms, specialized in design-intensive small-batch production, has grown steadily through the 1990s. Though largely un-researched, a recent survey suggests that there are now over 100 of these designer-maker firms in the Quarter (JQRP, 2002; see Figure 10.1).

Recasting the quarter

There are currently two strands of policy thinking regarding jewelry production in Birmingham: one concerns the kind of jewelry produced, the second concerns the physical fabric of the Quarter in the broader context of Birmingham's regeneration.

Jewelry becomes a "creative industry": the design focus

In line with my above comments about the state of the jewelry industry in Birmingham, a recent assessment of the U.K. industry by the Department of Trade and Industry noted U.K. producers' continuing loss of market share, lack of sector leadership and relative inattention to consumer needs and preferences. Key to the future of the industry, the report argued, was a stronger role for design:

> Designer-makers are the creative core of the jewelry sector. Their creativity and design ideas should be an input directly or indirectly into the mainstream manufacturing sector. Without this link and without a flourishing designer-maker sector, it is unlikely that the U.K. jewelry manufacturing sector will thrive or long term even survive.
>
> (DTI, 2001: 72)

In Birmingham, too, the steady decline of traditional jewelry manufacturing has led to a push to nurture designer-makers, viewing them as central to attempts to encourage the industry up market, using new materials and technologies (Birmingham City Council, 2001). For the West Midlands region the creative industries sectors, which include jewelry as a "crafts" industry, have been mapped onto different industrial clusters. In addition to this policy push from established agencies in the region, the Quarter's internationally renowned Jewelry School, part of the Birmingham Institute of Art and Design (BIAD) at the University of Central England, has played a pivotal role in training designer-makers and then supporting their transition into business.

The jewelry quarter as an "urban village"

The other main policy initiative related to jewelry production in Birmingham is a project designed to address the declining physical fabric of the Jewelry Quarter (see Pollard, 2004). The "Urban Village" project commenced in 1998 as an outcome of consultation between Birmingham City Council, English Partnerships (including the regional development agency, Advantage West Midlands), the Urban Villages Forum (now part of the Prince's Foundation) and British Waterways. The £250 million, five-year plan aimed to "transform the Jewelry Quarter into one of the U.K.'s most exciting mixed use areas where employment, including traditional jewelry manufacturing, residential, social and leisure facilities will all exist together" (JQUVPB, 1998: 3). One of the most significant objectives of the proposals was a ten-fold increase in the residential population in the Quarter to a target of 4,000–5,000 people, binding the Jewelry Quarter to the City Council's City Living Initiative. The proposals, confirmed in Supplementary Planning Guidance to the Birmingham Unitary Development Plan, relaxed planning restrictions around the Quarter in order to encourage investment and development. In addition to 2,000 new residential units, related achievements included the opening of a new gallery, some new cafes, shops and restaurants, improvements to pedestrian and educational pavement trails and mobile closed-circuit television cameras (JQRP, 2002).

Rather than focusing on industry specific networks, institutions or innovation systems, the Urban Village proposals sought to boost the industry by focusing on the Jewelry Quarter's physical links with the city center, its stock of residential premises and community and tourist facilities (see EDAW, 1998;

Pollard 2004). This approach has generated many tensions and highlighted the multiple, and not always complementary, agendas that are jostling for space in the name of supporting a creative-industry quarter in the city. In what follows I consider some of the jewelers' concerns with these specific policy proposals before considering some of the silences in these policies that need to be addressed if the industry is to push into higher value-added, more design-intensive markets.

Manufacturing culture? the materialities of jewelry production

In order to understand different groups of jewelers' problems with the creative industry and urban village proposals, it is important to understand the materiality – the raw materials and production processes and networks – of jewelry manufacturing. Firms in the Quarter are involved in a division of labor that, at its most rudimentary, can be broken down into three stages (Scott, 2000). The first stage often involves the cutting and preparation of a rough gemstone. This is followed by the design and production (casting, stamping or cutting) of precious and semi-precious metal settings and then the final setting of the stone in the metal casing. Birmingham's uniqueness stems from the extent of its elaborate social division of labor that has evolved around these three basic tasks. Around a core of manufacturing firms and designer-makers working predominantly with gold and silver are a host of small firms and outworkers specialized in stone and bullion dealing, refining, gem setting, die making, casting, finishing, stamping, piercing and spinning metal and in supplying specialist tools. Larger manufacturers employ perhaps 50–60 workers and make some use of outworkers for particular specialist tasks, such as casting or enameling. Smaller producers rely more extensively on outworkers, not only for castings, but also for gem settings, polishing and other stages of production. The Quarter is also home to a further supporting cast including importers and wholesalers, machinery and tool manufacturers, the Jewelry School and local trade associations.

The central problem regarding the Urban Village proposals, as far as manufacturers and designer-makers in the Quarter are concerned, is that in seeking to exploit the Jewelry Quarter as a heritage site for tourists, changes are under way which are threatening the economic viability of the very industry that has produced that heritage. These concerns take three specific forms.

The first concern of manufacturers and designers in the Quarter is the need for security. The raw materials of the trade – gold, silver, platinum and precious stones – mean that jewelers require secure premises and storage areas for finished and unfinished pieces. Moreover, the intricate social division of labor in the trade generates the need for outworkers, manufacturers and other trusted "runners" to transport unfinished pieces around the Quarter. For this reason, there is a degree of nervousness about the arrival of more "outsiders" and non-jewelry trades coming into the Quarter:

> We have enough security problems in the area already. I would much rather be in a street full of jewelry manufacturers than I would if it was other industries; if I walked out the door and got knocked over the head in a Jewelry Quarter people would know what was going on.
>
> (Manufacturer, source A)

A second concern for jewelers is the rapid increase in residential space in the Quarter. There are now over 1,000 residential units in the Quarter, housing over 1,200 residents (JQRP, 2002), and the original Urban Village proposals, if realized, will generate a residential population of 4,000–5,000. Jewelers located near residential units fear that their normal working practices, which often include late-night and weekend working at certain times of the year, may be affected by complaints from residents. Jewelers use a range of heavy stamping, pressing and milling machinery that generates noise and vibration. In addition, toxic materials, including cyanide and ammonia, are used to dip, clean and finish pieces of jewelry. Spillages, leaks and other accidents are a part of the trade: "This is what they didn't look at: the noise, the vibrations, the cyanide, the ammonia. Every now and then this street is blocked off because of another ammonia leak" (trade representative, source F).

In addition to fears that new apartments do not blend in well with the noise, vibration and emissions associated with jewelry manufacturing, the third and most significant fear of the jewelers is the effect of development pressure on land and property prices in the Quarter. Adjacent to the city center, the Jewelry Quarter has proved an attractive proposition for developers keen to build luxury apartments:

> I think there is also a very real element of fear about the flats particularly bumping up rates for workshops which are very cheap for most people, or they will share the rent. So there is a very real worry about that. And to be honest, I have seen the prices rocket and its developers moving in and I actually wonder at the wisdom of it.
>
> (Educator, source J)

Many of the apartments being built in the Quarter, like those in the nearby city center, were priced in the £300,000–£400,000 bracket (in 2002) and aimed at investors and professionals from outside the area (Gray, cited in Arnot, 2002).

Rapidly increasing land and property prices are also pushing established manufacturers with large premises to contemplate their futures:

> I think the future at the moment is, if residential prices around here go at the rate they appear to be talking about, then we are all absolute ass's sitting here with large factories. We might as well turn them into flats. I mean I think it's quite astonishing. There's a lot of properties around here, it will dawn on people that the property is worth more than their businesses.
>
> (Manufacturer, source A)

For this, often moderately wealthy, constituency of producers a commitment to manufacturing jewelry or preserving the family business to pass on to children fuels their decision to stay in the Quarter. Another manufacturer commented:

> I am very wary. I am old enough to remember that we had a Gun Quarter once. It was very similar. It was the same high craft input, low wages, low overheads. . . . And with the gun quarter they said "Let's tidy it all up. We'll pull it all down and put you a new factory up and, by the way, rent is going to be 10 times more than what you pay now." And all these guys said "No, I'll go and work on the track at Longbridge [car factory], I'll earn more money, less stress."
>
> (Manufacturer, source B)

In the Spring of 2001 a survey of 118 businesses in the Quarter (Anonymous 2001) cited reported manufacturers' fears regarding excessive residential development, rising property prices, conflicts between residential and commercial uses and the consequent loss of complementary businesses. Birmingham City Council too were well aware of the "pressure for inappropriate residential development in the industrial heart of the Quarter, threatening the interdependent and increasingly fragile structure of the jewelry trade" (2001: 29). The City Council, English Heritage and Advantage West Midlands commissioned a "Conservation Area Character and Management Plan" for the Jewelry Quarter Conservation Area in 2001 in recognition of these pressures. In January 2002, the City Council adopted new Supplementary Planning Guidelines (Anonymous, 2002) designating the Jewelry Quarter as a Conservation Area and, crucially, limiting further residential development in the inner industrial core and retail area.

Manufacturing culture? Material, cultural and social assets in the quarter

In some policy circles in Birmingham there is acknowledgment that the jewelry industry has been neglected compared with the higher profile and more vocal automotive and engineering industries in the region (policy maker, source C). In this context, whatever the unintended consequences of policy, it could be argued that the Urban Village and Creative Industry proposals are at least a starting point for the industry. Nevertheless, there are other obstacles to progress in the Quarter that are not directly being addressed by policies focused on the cultural credentials of the Quarter and its products.

One of the major predicaments reported by manufacturing jewelers who are keen to survive by producing exceptionally high-quality jewelry in the Quarter is

> an indifference, a terrible indifference, to quality as compared with 30 years ago. Amazingly depressing. The biggest change in the U.K. is, starting about 25 years ago, was the expansion of a leading London retailer who

> bought up county jewelers who were our mainstay . . . there was hardly a county town in the country that didn't lose a good family jeweler. . . . And the U.K. now, you will hear the same story from [another manufacturer], you can count on two hands the real quality retailers outside of the West End [of London].
>
> (Manufacturer, source A)

With the arrival of mass-production techniques in jewelry, there also arrived a consolidation of jewelry retail chains in the U.K. Another manufacturer argued that retailers were responsible for encouraging a culture of price undercutting in the Quarter by playing off different producers against each other (manufacturer, source W). More recently, non-specialist retailers, including clothing stores and chemists, have started selling cheap fashion jewelry which is usually silver or nine-carat gold (Key Note, 1998).

The lack of quality distribution outlets is felt acutely in the Quarter. One of the major changes in the Quarter since the 1970s has been the introduction of jewelry retailers. The first retailers arrived in the late 1970s and their numbers increased sharply in the recession of the early 1980s.[4] There are now approximately 150 jewelry retailers in the Quarter (JQRP, 2002), yet just three outlets are known to sell locally designed and made jewelry (designer-maker, source G). Moreover, designer-makers and manufacturers alike are dismissive of the quality of the, largely imported (EDAW, 1998), jewelry sold in the Quarter. While retailers are keen to see Heritage Trails and greater foot traffic in the area, most manufacturers and designer-makers currently have little or no direct dealings with the public and have little to gain from increased foot traffic if their work is not being displayed locally.

A second, and increasingly serious, problem in the Quarter concerns the reproduction of skills. The jewelry industry is known as a relatively low-wage industry that has traditionally relied on apprenticeship schemes to train the highly specialized craft workers required for the hand-made part of the market. In Birmingham, there are growing concerns about the lack of young people entering the industry:

> . . . if the British jewelry industry is going to keep going then you need that base of just the skilled workforce. You're finding that kids aren't coming in like they used to be, the kids coming in now, the youngest ones I know are in their late 20s.
>
> (Designer-maker, source O)

The intensely difficult market conditions for jewelers since the early 1990s have also, to the chagrin of local policy makers, eroded jewelers' commitments to apprenticeship schemes. As one manufacturer put it, "There's not a lot of apprenticeships because you've got to remember this is not the time to carry passengers" (manufacturer, source B). The same manufacturer continued:

> My engraver who does inscriptions on wedding rings, you know signet rings, whatever, he is over 60. The chap who does this sort of work for us engraving on stone and gold, he's 70. My setter is about 50 odd. My one enameler who I give my really best work to, he is 70 . . . and the chap that sets my signet rings . . . I don't know how old he is but he looks about 150 [laughs]. He works at home. There is nobody following him . . . there is nobody I know in the trade who has that skill at their fingertips, for me to say that I know that it's going to come back looking perfect.

Concerns about the inability to produce traditional skills in the Quarter do not just affect the high-end manufacturing jewelers, but also the designer-maker community that has been the focus of policy attention. Designer-makers are typically trained in a range of jewelry skills, rather than specializing in one part of the trade, and will, depending on the kinds of materials they work with, make use of specialist outworkers in the Quarter. Some designer-makers perform most tasks themselves and use outworkers solely for casting work; others simply produce drawings of their designs and outsource every aspect of production from model making and casting, to gem setting, enameling, polishing and other kinds of work. Forty of 48 designer-makers surveyed in the Quarter in 1999 used subcontractors to make parts of their work (Taylor Burgess Consulting Limited, 1999). One designer-maker observed:

> The designer-makers can't do everything themselves. They rely on all the little workshops to go and get their casting done, to get etchings done, whatever, so it's a mutual skills base . . . It's all very well supporting the designer-makers but without the skill base to back that up, the designer-makers are dead on their feet.
>
> (Designer-maker, source O)

Although the Jewelry School occupies a prominent position in policy debates in the Quarter, it has been criticized by manufacturers for its focus on producing graduates with strong design, as opposed to manufacturing, credentials:

> I don't think the Jewelry School produce the right sort of people: I don't think it's terribly clever just producing designer jewelers. I think they should have people learn, say, to be an engraver, learn to be a diamond setter, not to have a little play at that and a bit of engraving and a bit of enameling. I don't think they really push people to become highly skilled.
>
> (Manufacturer, source B)

Criticisms of the School and its emphasis on design reveal different conceptions of "skill" in the Quarter and also different perceptions about the future viability of the manufacturing-led model of jewelry production that has been handed down, often staying in particular families, since the eighteenth century.

It is here that a third, and oft-cited problem of the industry rears its head. As policy makers, manufacturing and designer-maker constituencies will acknowledge, the Quarter houses some very hierarchical social relations and ingrained understandings and cultures of working that have done little to boost recruitment or to help the image of the industry. To generalize, the jewelry industry is notable for its high degree of fragmentation – it is dominated by large numbers of small, often family-run firms – and a tradition of insularity, conservatism and passivity (sources, manufacturer A, E, F; policy maker P). A climate of secrecy may be a product of the large numbers of family-run businesses in the Quarter and the understandable security concerns of jewelry production. Nevertheless, manufacturers have suggested that the ongoing decline in the trade is eroding any remaining culture of co-operation: "Oh it's dog eat dog, definitely, in the last ten years or so. I think it is because of the imports" (manufacturer, source W).

Designer-makers, in addition to having difficulty finding appropriate outlets for their work, have also experienced difficulties working with manufacturers:

> [designer-makers] don't want to rip off other people because it's like, unfortunately, that's what big companies do. They come round the shows, they send people round to pick up ideas and you see them mass produced, it's cheap and that's the worry.
>
> (Designer-maker, source X).

Another designer-maker commented:

> I know when I came to Birmingham we were all moaning about the older jewelers here that wouldn't tell you anything. I mean what have they got to lose? We're not going to steal their customers. We're totally different to them. And they just didn't want to tell us anything, where to go to get things, techniques or anything.
>
> (Designer-maker, source P)

Age is but one axis of difference between manufacturers and designer-makers. The sensitivities around the import being given to design are also testimony to broader demographic changes ongoing in the Quarter that challenge traditional cultures and understandings of work:

> They are resented a bit I think, the designer-makers. I think they [the manufacturing jewelers] think they are upstarts, because they're not "in" the trade, they're not outworkers, they're not just taking in jobbing work, they're going out and doing their own thing. There's people like [a designer-maker] over the road who does designer-maker jewelry but it is precious metal, it is precious stones, so he is respected, he is OK. Anybody who's not doing precious metals, or it's more fun and funky or whatever, it's "ooh, it's not jewelry."
>
> (Designer-maker, source O)

Another designer-maker spoke about the increasing diversity in the Quarter, not only in conceptions of skill and employment relations but also in gender and educational background: "They're [manufacturing jewelers] not too happy about it [the constituency of designer-makers] being women, and they're not too happy about the fact that we went to university, college, or whatever" (designer-maker, source P). In the absence of any systematic survey of the demographic make-up of designer-makers in the Quarter, evidence is largely anecdotal; various sources suggest, however that at least 75 percent of all designer-makers are women (Taylor Burgess Consulting Limited, 1999; designer-makers sources G, O, P). While white and Asian women have graduated from the Jewelry School to set up their own designer-maker businesses, policy makers seeking to encourage manufacturing jewelers in the Quarter to draw new recruits from surrounding areas of high unemployment have found:

> locals won't work in the industry because it's seen as predominantly white and male, and they [the manufacturing jewelers] wouldn't recruit people from there because they might be untrustworthy, because they come from an area that is a bit suspect or whatever. We had preliminary discussions about getting local people skills for jewelry, but it never happened.
>
> (Policy maker, source S)

To summarize, in addition to the problems highlighted via jewelers' reactions to specific policy initiatives designed to rejuvenate the industry and the Quarter, there are a number of other concerns voiced by manufacturing and designer-maker jewelers keen to occupy high-value markets. One legacy of the post-war turn to mass production is a lack of appropriate distribution networks for high-quality, design-intensive jewelry in Birmingham. The industry also faces persistent difficulties in reproducing skills. Finally, as the industry becomes more design-conscious, what is valued as a "skill" is now contested. In a climate of mistrust, conservatism, fear and decline, traditional manufacturing cultures – ways of doing business – in the Quarter are facing a series of economic, social, cultural and demographic "shocks" en route to building the capacity to survive and push into higher value-added, more design-intensive jewelry markets.

Conclusion

In the context of Birmingham's ongoing reinvention and the central role of cultural production and consumption, this chapter has examined two attempts to reinvigorate one of the city's oldest manufacturing districts, its Jewelry Quarter: first, by promoting it as a modern, design-led or "creative" industry; and second, by expanding its residential and leisure facilities to exploit any tourist potential. These policy initiatives have undoubtedly improved some aspects of the physical infrastructure of the Quarter and have also helped to increase the visibility of its design community. Yet they also expose some of the divergent, if not contradictory, agendas being played out in the Quarter in the name of promoting the city's "creative industries." Beyond questioning what, precisely, policy makers in Birmingham want to construct, this case study illustrates the diversity and complexity of the idea of "creative" industries.

There are three other points that can be drawn from this study. First, and as a means of grasping some of the diversity and complexity I allude to above, I have argued that it is instructive to examine the experiences of firm owners and workers who work with and live through such policy initiatives. The Urban Village proposals bound the Jewelry Quarter into the economic logic of other regeneration schemes that prioritize cultural production and consumption in order to attract investors and affluent consumers to Birmingham. Yet this gentrification has already put upward pressure on land and workshop prices and priced some jewelers out of the Quarter. Although the recent tightening of planning regulations to preserve industrial premises in the Quarter has ensured the short- to medium-term future of many jewelers – who would otherwise be looking to a downturn in local property markets to dampen speculation – it raises the question of who is included and excluded in such regeneration schemes and whether there is space for a relatively low-wage, cash-poor, craft-based industry in contemporary imaginaries of Birmingham as a "capital of culture."

Second, and related, policies aimed at promoting the cultural and creative credentials of Birmingham's jewelers – considerable as those credentials are – cannot ignore the materiality of jewelry production. Designers in the Quarter draw on an intricate, extended social division of labor to transform their designs into finished articles. The Urban Village model of regeneration in particular fails to acknowledge the daily material realities of production – the flows of people, machinery and materials –which create jewelry.

Finally, these policies, framed by a growing interest in "culture," are interesting for their silences on various aspects of manufacturing culture in the Quarter. Birmingham's Jewelry Quarter is not populated with a homogeneous manufacturing and design community who share a common project or vision. When examining the constituencies of manufacturing jewelers and designer-makers, cleavages of age, class, race and gender are just some of the dimensions of difference on display that construct varying definitions of skill, different traditions and understandings of training and recruitment and different valuations of the worth of different kinds of jewelry. The differences between the old and the new are only sharpened by the intensely competitive conditions facing the industry, conditions that have eroded commitments to the reproduction of skills, reduced collaboration and trust and led to destructive price competition in some markets.

Birmingham's future in jewelry production lies in pushing into high-quality manufacturing and innovative, design-intensive jewelry. In this endeavor, there are some promising signs; the city has a long legacy of skilled metal-working and is an established U.K. center and something of a magnet for young designers. Nevertheless, the evolution to a competitive, more design-intensive jewelry industry requires a closer integration than currently exists between manufacturing and design. It would also involve the re-building of the trust, collaboration and strategic capacity necessary to thrive in higher value-added markets. Some aspects of these requirements could be addressed by public policy to encourage the development of retail space for locally designed and produced jewelry, and to forge links between jewelers and other designers in the region working with textiles, furniture and other craft goods.

Yet, there are other aspects of change which are likely to be much more difficult and elusive. Moving into higher value-added markets, maintaining and increasing skill levels and forging links with other design-intensive industries may be the only way for jewelers to secure their economic future, yet for many that challenges "the value of material and social assets, identities, commitments, and social power . . . tied up in a particular constellation of material practices, social relations and ways of thinking" (Schoenberger, 1997: 227). This is arguably the most imposing cultural challenge facing Birmingham's communities of jewelers, trade bodies and policy makers keen to build on the history and the material and symbolic importance of the region's jewelry manufacturing prowess.

Acknowledgment

This research was undertaken as part of a broader Nuffield-funded project (grant number SGS/0495/G).

Notes

1 Hallmarking figures can be difficult to interpret. Not all material hallmarked in Birmingham has necessarily been made locally; there may be significant delays between production and hallmarking and jewelry made of non-precious metals is not subject to hallmarking (Mason, 1998).

2 The database of firms used to construct the map are taken from unpublished Business Link database (for 2001), augmented by the author's own survey. The classifications used are more conservative than those used by Wise (1949). For example, manufacturers of optical goods, bead, bolt and button makers and factors, and wholesalers are excluded.

3 The Birmingham Jewellers and Silversmiths Association became the British Joint Association of Goldsmiths, Silversmiths, Horological and Kindred Trades (abbreviated to BJA) in 1946 (Mason, 1998).

4 It is unclear whether this expansion was a product of other retailers moving in the Quarter, or manufacturers or redundant workers seeking to generate more cash flow (Smith, 1987).

References

Anonymous (2001) "Jewelry Quarter business survey," *Hockley Flyer*, 192: 35.

Anonymous (2002) "Planning matters," *Hockley Flyer*, 200: 50.

Arnot, C. (2002) "Gem snatch," *Guardian*, 20 February 2002.

Barnes, T. (1996) "Logics of dislocation: models, metaphors and meaning in economic geography," New York: Guilford Press.

Bianchini, F (1993) "Remaking European cities: the role of cultural politics," in F. Bianchini and M. Parkinson (eds) *Cultural Policy and Urban Regeneration: The West European Experience*, Manchester: Manchester University Press.

Birmingham City Council (2001) *The Jewelry Quarter Conservation Area Character Appraisal and Management Plan*, Birmingham: Birmingham City Council.

Crang, P. (1997) "Cultural turns and the (re)constitution of economic geography," in R. Lee and J. Wills (eds) *Geographies of Economies*, London: Arnold.

Crewe, L., Gregson, N. and Brooks, K. (2003) "Alternative retail spaces," in A. Leyshon, R. Lee and C. Williams (eds) *Alternative Economic Spaces*, London: Sage.

Department of Culture Media and Sport (1998) *The Creative Industries Mapping Document*, London: DCMS.

Department of Trade and Industry (2001) *The Competitiveness Analysis of the U.K. Jewelry Sector*, London: DTI.

Devereaux, M. P., Griffith, R. and Simpson, H. (1999) *The Geographic Distribution of Production Activity in the U.K.*, London: Institute for Fiscal Studies.

EDAW (1998) "The Jewelry Quarter Urban Village, Birmingham," Birmingham: EDAW.

Florida, R. (2002) "Bohemia and economic geography," *Journal of Economic Geography* 2: 55–69.

Gdaniec, C. (2000) "Cultural industries, information technology and the regeneration of post-industrial urban landscapes: Poblenou in Barcelona – a virtual city?," *Geojournal*, 50: 379–87.

Hall, P. (2000) "Creative cities and economic development," *Urban Studies*, 37: 639–50.

Henry, N., McEwan, C. and Pollard, J. S. (2002) "Globalization from below: Birmingham – postcolonial workshop of the world?," *Area*, 34: 118–27.

Hubbard, P. (1996) "Urban design and city regeneration: social representations of entrepreneurial landscapes," *Urban Studies*, 33: 1141–461.

Jewelry Quarter Regeneration Partnership (2002) *The Business Plan 2002–3*, Birmingham: Jewelry Quarter Regeneration Partnership.

Jewelry Quarter Urban Village Partnership Board (1998) *Jewelry Quarter Urban Village Prospectus*, Birmingham: JQUVPB.

Key Note (1998) *Jewelry, Watches and Fashion Accessories*, London: Key Note Ltd.

Kong, L. (2000) "Cultural policy in Singapore: negotiating economic and socio-cultural agendas," *Geoforum*, 31: 409–24.

Lee, R. and Wills, J. (eds) (1997) *Geographies of Economies*, London: Arnold.

Loftman, P. and Nevin, B. (1996) "Prestige urban regeneration projects: socio-economic impacts," in A. Gerrard, T. Slater and R. Studley (eds) *Managing a Conurbation: Birmingham and its Region*, Studley: Brewin Books.

Mason, S. (1998) *Jewelry Making in Birmingham 1750–1995*, Chichester: Phillimore and Co. Ltd.

Norwood, G. (2002) "Safe houses? Not any more. With many homes empty, buy-to-let can be tricky," *Observer*, 10 March 2002.

Peet, R. (1997) "The cultural production of economic forms," in R. Lee and J. Wills (eds) *Geographies of Economies*, London: Arnold.

Pollard, J. S. (2004) "From industrial district to urban village? Manufacturing, money and consumption in Birmingham's Jewelry Quarter," *Urban Studies*, 41: 173–93.

Roche, J. (1927) *The History, Development and Organisation of the Birmingham Jewelry and Allied Trades*, Birmingham: Dennison Watch Case Company Ltd.

Sayer, A. (2001) "For a cultural political economy," *Antipode*, 33: 687–708.
Schoenberger, E. (1997) *The Cultural Crisis of the Firm*, Cambridge, MA: Blackwell.
Scott, A. J. (2000) *The Cultural Economy of Cities*, London: Sage.
—— (2001) "Capitalism, cities, and the production of symbolic forms," *Transactions of the Institute of British Geographers*, 26: 11–23.
Smith, B. M. (1987) *Report of an Interview-Based Survey in the Birmingham Jewelry Quarter Relating to the Jewelry Industry*, Birmingham: Centre for Urban and Regional Studies, University of Birmingham.
Storper, M. (1997) "The city: centre of economic reflexivity," *Service Industries Journal*, 17: 1–27.
Taylor Burgess Consulting Limited (1999) *Designer/Maker Study within the Jewelry Quarter, Birmingham*, Birmingham: Taylor Burgess Consulting Ltd.
Thrift, N. and Olds, K. (1996) "Refiguring the economic in economic geography," *Progress in Human Geography*, 20: 311–37.
Ward, K. (2001) "Doing regeneration: evidence from England's second cities," *Soundings: A Journal of Politics and Culture*, 17: 162–6.
Webster, F. (2001) "Re-inventing place: Birmingham as an information city?," *City*, 5: 27–46.
Wise, M. (1949) "On the evolution of the jewelry and gun quarters in Birmingham," *Transactions of the Institute of British Geographers*, 15: 59–72.

11 Beyond production clusters

Towards a critical political economy of networks in the film and television industries

Neil Coe and Jennifer Johns

Introduction

There is a long-standing theoretical and policy preoccupation with notions such as "industrial districts" and "clusters" when describing the governance and spatiality of networks in the film and television industries.[1] In this chapter, we assert that such analyses run the danger of over-privileging the importance of local institutional and organizational network relations, and thereby downplay the significance of a range of extra-local networks upon which the nature, or indeed the very existence, of these formations may depend. Instead, we argue for a conceptualization that recognizes how the majority of so-called film industry clusters or industrial districts are embedded in, and shaped by, a complex web of multi-scalar network connections. Effective analysis and ensuing policy prescriptions must recognize the key extra-local network relations – be they intra-regional, intra-national or international – upon which agglomerations are predicated. In short, we need to develop a non-deterministic "critical political economy" of these networks that considers exactly where the power to initiate, foster and develop local network formations resides, how it is enacted, and what its implications are. This in turn necessitates a broad focus that looks beyond the actual film ***production*** process to reveal the key domains of finance, distribution and exhibition that exert a massive influence on the organizational structures and geography of the industry.

In this chapter we use empirical evidence from agglomerations within the film industries of Canada and the U.K. – more specifically Vancouver and Manchester – to demonstrate the limitations of existing conceptualizations. The following sections will review existing research into the organizational structure and geography of the film industry and outline our conceptualization the film production system, highlighting the key network connections within which power and influence reside. The global market for the film industry is estimated to be worth $60 billion annually (U.K. Film Council, 2003), but ownership and control of this industry is increasingly concentrated in Hollywood. Our case studies will demonstrate how localized centers of production can be conceptualized to incorporate connections to global networks of finance and distribution.

Beyond production clusters: reassessing governance and territoriality in the film-production system

In this section we critically evaluate existing research into the film industry. Two lines of argument underpin our review. First, we identify a preoccupation in the literature with film *production* processes which obscures the key functions of finance, distribution and exhibition and thus the power or governance structures that shape the nature of the industry. Second, this narrow focus on production leads to a weak understanding of the territoriality of the industry by over-emphasizing the importance of localized production clusters at the expense of key extra-local connections to financiers and distributors.

Let us first consider research into organizational restructuring and governance within the film industry. There has been surprisingly little recent research on the film industry by social scientists given its overall economic significance. What research there has been reveals the lasting influence of the work of Michael Storper and Susan Christopherson who – some fifteen years ago now – presented their interpretation of the post-war restructuring of the U.S. motion-picture industry as part of an emerging broader discourse of "flexible specialization" (Christopherson and Storper, 1986, 1989; Storper, 1989; Storper and Christopherson, 1987). In their account, by the 1970s, the vertically integrated Hollywood studio system – with most production functions retained in-house – had been replaced by a vertically disintegrated production system characterized by a flexible mass of smaller production and service entities. As a result, "the production process is no longer carried out within the firm, but instead has moved to the external market, carried out through a series of transactions linking firms and individuals in production projects" (Storper and Christopherson, 1987: 107).

Storper and Christopherson perhaps presented the flexible specialization of the film industry in rather favorable terms. They described a situation in which the role of large firms in direct production diminished over time, to be replaced by a larger number of more diverse firms. Yet, they also maintained that a hierarchy of control remained as the major studios continued to dominate financing and distribution. This ambiguity is one of the main sources of debate surrounding their work. In emphasizing the importance of distribution, Aksoy and Robins (1992) argued that oligopolistic control never ceased to be a distinguishing feature of Hollywood. They suggested that from the 1950s onwards, although more production was conducted by independents, the key point was that they were increasingly being financed and distributed by the major studios:

> By holding on to their power as national and international distribution networks, the majors were able to use their financial muscle to dominate the film business and to squeeze or use the independent production companies. Independent production was used to feed the global distribution networks that the majors had built.
>
> (Aksoy and Robins, 1992: 9)

Hence their key criticism of Storper and Christopherson's account lay in the apparent over-emphasis on the production phase, and their relative neglect of finance, distribution and exhibition – all key aspects of the film-production system.

Despite this shortcoming, the flexible specialization thesis has been also been applied to the television industry in the U.K. Barnatt and Starkey (1994) describe the disintegration of large organizations in the television industry, resulting in an increased differentiation of small, specialist service providers and the replacement of the rigidities of vertically integrated relationships by external contracting. They examine the U.K. television industry as a form of "dynamic network," which is "at once flexible and specialized" (Barnatt and Starkey, 1994: 252). The flexible networks are predicated upon the creation of four groups of flexible agents: specialist freelancers, performing artists, facilities houses and contract services. These groups are viewed as forming the nodes around a production core and may be either internal or external in nature. Each node has a particular specialization, and as multiple organizations become involved in a network, the boundaries between individual firms become blurred.

Barnatt and Starkey's work represents an attempt to understand the increasing complexity of inputs required for the television production process. Their "flexible network" concept clearly has its merits: the adoption of a "network" discourse contributes to a better understanding of the industry, which essentially operates through networked organizational forms that draw together individuals and firms via contractual arrangements as and when necessary but, as with the work of Storper and Christopherson, it can be accused of being overly romanticized. In particular, due to a focus on TV *production*, there is little consideration of the inherent power relations driving the construction and maintenance of such networks. Tempest *et al.* (1997) argue that the search for a dominant production paradigm itself is counter-productive, arguing that the U.K. television industry demonstrates a diversity of different production processes. Significantly, they suggest that whilst "dynamic networks" may make up the production teams, it is the organizational context in which these networks operate that determines the nature of the production system as a whole. This bigger picture cannot be gained by a narrow focus on small firm interaction, particularly as the media industries are increasingly being dominated by national and international concerns.

These debates demonstrate the dangers of adopting too narrow a focus on production, rather than the production system as a whole. The flexible specialization approach struggles to deal with the complexity of interrelationships between the functions of finance, production, distribution and exhibition, and the companies performing them (Blair and Rainnie, 2000). But who, then, controls these other key functions in the production system? In short, an increasing share of film production is controlled by a small group of large global organizations with international interests in a wide array of media forms. The 1990s witnessed an unprecedented series of mergers and acquisitions among global media giants, resulting in the emergence of a "tiered global media market"

(Herman and McChesney, 1997). Ten enormous conglomerates or "majors" form the first tier. In 2000 these were: AOL/Time Warner, Vivendi Universal, Disney, Bertlesmann, Viacom, News Corporation, Tele-communications Inc (TCI), Sony, Philips/Polygram and General Electric/NBC (Louw, 2001). Of these, News Corporation, AOL/Time Warner, Disney, Bertlesmann, Viacom and TCI can be described as vertically integrated media conglomerates as they are major producers of entertainment and media software *and* have global distribution networks. The second tier of global media organization consists of approximately three dozen fairly large firms that fill regional niches within the global system (Mermigas, 1996). These firms do not operate independently – they all are involved in joint ventures or strategic alliances with one of the giant conglomerates. The third tier is composed of many smaller national and local firms, again filling certain niches and providing services to the larger firms (Herman and McChesney, 1997).

Over time, the majors have consolidated and further integrated their operations, growing in size as a result. They have become key actors or "gatekeepers" in the industry through their control of distribution networks, and as a result have much more flexibility over financing and production of their own projects. As the media conglomerates have acquired assets over an increasing range of media, their power over distribution has also stretched. A crucial point is that despite the continued existence of localized production centers, few are able to exist without connection to the distribution networks controlled by the giant conglomerates:

> The forces shaping the industry remain beyond the control and influence of local actors. The agenda is shaped by the activities of large media companies, and local or regional production industries can only struggle to adapt to the adverse conditions of this environment.
>
> (Robins and Cornford, 1994: 235).

Such debates lead us on to considering the inherent spatiality of the film-production system. According to Storper and Christopherson, the new flexibly specialized system exhibited what they termed a "split locational pattern." On the one hand, tight transactional networks created strong agglomerative forces that ensured that the bulk of studios, production houses and subcontractors remained in the Los Angeles area. On the other hand, the newly flexible system also lent itself to the shooting of productions in locations outside Los Angeles, and increasingly, outside the USA. Much of their analysis, however, focused on the dynamics within the Los Angeles agglomeration and labor market, rather than on film production centers that have developed through location shooting. Similarly, Scott's recent work on the film industry (2000a, 2000b) is more concerned with how the primacy of certain agglomerations is sustained, rather than how new centers of production and innovation may emerge. Recent developments make this research lacuna more puzzling, as the geography of film production has expanded to incorporate an ever-wider range of locations across the globe.

Contemporary research into the film industry – and indeed the cultural industries more generally – makes great play of the *local* dynamics of growth within agglomerations of activity. This perspective favors examination of local networks, to the relative neglect of other "longer" network linkages at the national and global scale. The bias towards film *production* exacerbates this tendency by drawing attention to precisely those network relations that do depend upon proximity. A number of recent studies have begun to tackle the issue of how to situate these production clusters or agglomerations within national and global networks (e.g. Coe, 2000a, 2000b, 2001; Bathelt, 2002; Krätke, 2002). Often, as described in the quote by Robins and Cornford earlier, these extra-local networks will enable the funding and distribution of films upon which the production cluster depends. For example, Basset *et al.*'s (2002) study of natural history filmmaking in Bristol reveals many of the flexible dynamics of learning and innovation that the literature would lead us to expect. Significantly, however, they stress the critical role of U.S. satellite channels in providing a distribution and exhibition outlet for the end products. A greater range of studies are needed that reveal more fully the nature, extent and importance of the extra-local networks in which all productions clusters are embedded, to a greater or lesser extent.

The film production system: towards a critical political economy of networks?

We can develop our arguments further through a consideration of the film industry's production system. In our conceptualization, the range of inter- and intra-organizational network relationships shown in Figure 11.1 constitutes the production system. Following Storper and Harrison (1991), production systems can be characterized in terms of their structure, governance and territoriality. In basic terms, the *structure* of the film production system is simple, and can be split into six sequential phases. In the first, the rights to an idea or story are acquired, funds are raised and certain key individuals (director, lead actors) may be contracted to the project. As our feedback arrow in Figure 11.1 suggests, securing distribution contracts in advance is a common method of raising finance. The next three phases, pre-production, production and post-production refer to the actual "production" of the film. The pre-production phase covers everything that happens once a production is "green-lighted" until actual shooting starts, encompassing a wide variety of budgeting, scripting, planning and scheduling activities. The production phase is the most capital and labor intensive, as the huge range of elements needed to make a film is brought together in a studio or on location. These elements include the following: above the line (creative) costs such as stars actors and producers, and below the line (technical) costs such as studio rental, technical labor and a variety of other support services. Post-production covers the activities that transform the daily film footage into the finished product which can then be distributed for theatrical release or broadcast for television, such as editing and soundtrack

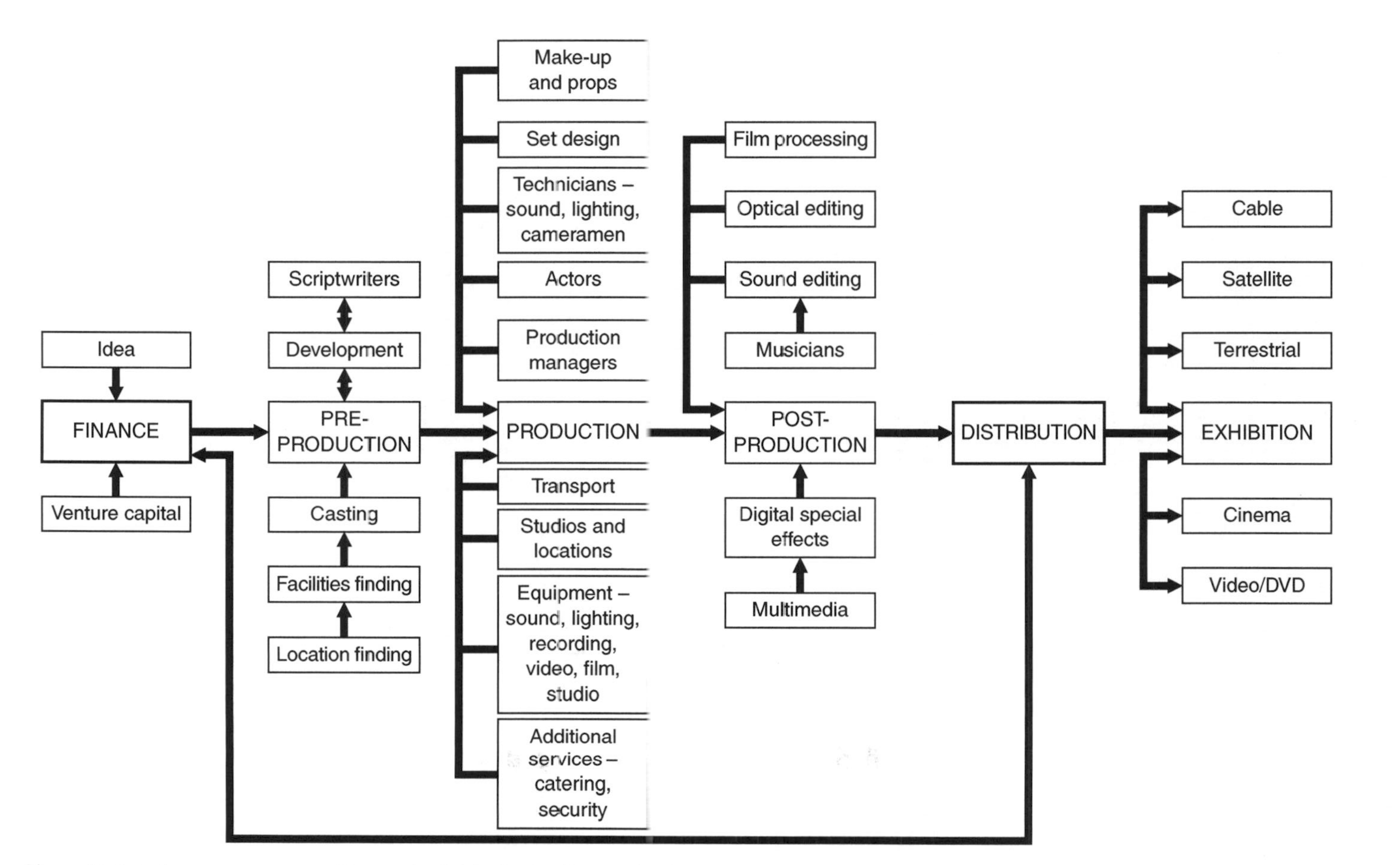

Figure 11.1 A six-stage sequence of inputs/activities in the film production system.

development. In the fifth phase, the finished project is promoted through various advertising media, and distributed, increasingly both domestically and abroad. Finally, the product will be exhibited through the appropriate outlet(s) (cinema, TV network, cable TV, video, etc.).

This production system is highly flexible in nature. Film and television productions are effectively short-term coalitions of directors, actors, crew and various service subcontractors, with each element being contracted separately to the project (see Grabher, 2002, for more on similar forms of working in other sectors). The length of the contract may range from four weeks for a television movie, anywhere from two to four months for the average theatrical release, to ten months for a television series of perhaps twenty episodes. Each element typically requires a contract negotiated between an individual, a union or a company, and the production entity. While a number of intermediaries are involved as coordinating agents in this process, it is usually the producer who puts together, and keeps together, the coalition for each project.

Within this flexible production system, the "majors" remain the ultimate contractor, while the various elements of pre-production, production and post-production are enacted by a coalition of small firms and union employees contracted to a temporary production entity. Although Storper and Christopherson are today still correct in their assertion that the premier agglomeration of such companies and labor is in Los Angeles, the flexible nature of the production system does offer opportunities to locations that can attract a considerable volume of production activity. This process is shifting the *territoriality* of the production system towards a more dispersed pattern. In the so-called runaway phenomenon, the phases of production where there is most expenditure are spatially dislocated from Los Angeles (or other control centers such as London or New York) and undertaken in lower cost localities. The traditional model has been for some pre-production tasks to be performed on location, along with some or all of the production process, and the most basic post-production activity, such as transferring the daily film footage onto tape. While these activities have proven to be spatially transferable, centers such as Los Angeles still retain a stranglehold on securing story rights, financing productions, undertaking post-production work, and controlling the networks of film distribution and exhibition. Thus, *governance* of the production system is still enacted from a small select group of cities.

Our notion of a "critical political economy" of networks explicitly recognizes that not all networks are equal. It is not enough to simply identify the various connections shown in Figure 11.1. Instead, we draw attention to the key finance and distribution relations: these two stages are closely linked as finance is often secured by selling distribution rights in advance of production. Without these key network connections in place, the rest of the production system cannot exist. In essence, power within the system largely resides with those that have the resources to finance and distribute films. In addition, the network connections in Figure 11.1 also need to be mapped out in space. Our case studies will reveal that in most production centers, while film production is enacted through flexible, localized small-firm networks, it is precisely the key finance and distribution connections that are most likely to be extra-local in nature. Such centers are thus dependent on external financial and creative decisions, and short-term inward capital flows. We use the term "critical" simply to denote a need to think carefully about the local economic development issues that arise from these conditions.

Case Study 1: runaway Hollywood production and the rise of the Vancouver film industry

The development of Vancouver as a film-production centre is intimately related to the post-war organizational restructuring of the Hollywood studio system and the concomitant rise of location shooting on a large scale (see Christopherson and Storper, 1986 for a review of these processes). Vancouver has been the single biggest beneficiary of the "runaway" production phenomenon – whereby Hollywood productions are filmed outside Los Angeles to achieve cost savings – with over 80 percent of its now considerable activity being attributable to these spatially mobile projects (see Coe, 2000a, 2000b, 2001 for an extended consideration of the Vancouver case).

Although filmmaking in British Columbia dates back to the late nineteenth century (Browne, 1979), a significant film industry has only really emerged in the province since the late 1960s, with the most important phase of growth occurring since the early 1980s (see Figures 11.2 and 11.3). Production grew steadily throughout the 1980s, reaching a level whereby 61 projects generated

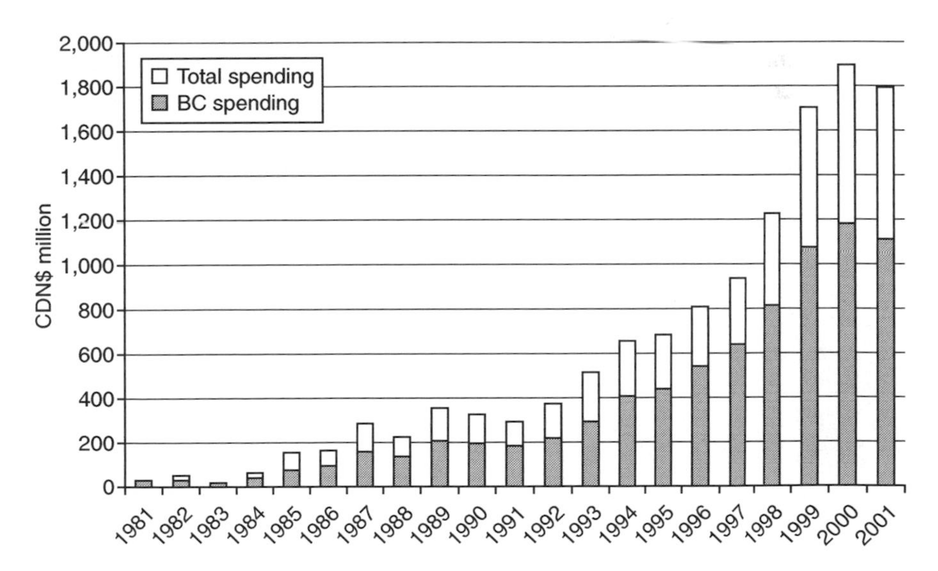

Figure 11.2 Vancouver film industry by value, 1981–2001 (source: http://www.bcfilmcommission.com).

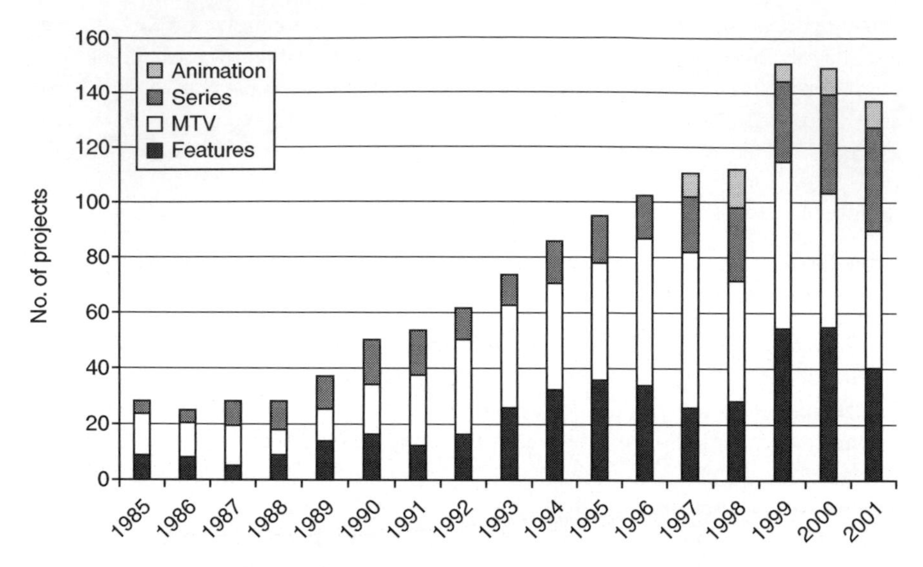

Figure 11.3 Vancouver film industry by no. of projects, 1985–2001 (source: http://www.bcfilmcommission.com).

CDN$211 million of revenue in the province from total budgets of CDN$368 million in 1992. Growth has accelerated in the 1990s as Hollywood runaway production has increased in response to the insatiable demand for low-cost programming from American cable television companies and newly independent television broadcasters. Table 11.1 summarizes the key parameters of the contemporary Vancouver film industry. In 2001, the industry was worth some CDN$1.1 billion to the British Columbian economy, from total spending of nearly CDN$1.8 billion, and supported an estimated 50,000 jobs. The data illustrates how the Vancouver sector is actually dominated by television production, which accounted for 65 percent of spending in 2001. Nineteen U.S.-financed television series alone accounted for 29 percent of total investment,

Table 11.1 Breakdown of spending in BC film industry in 2001

Genre	*Canadian (CDN$ millions)*	*No.*	*Foreign (CDN$ millions)*	*No.*	*Total (CDN$ millions)*	*No.*
Features	60.4	18	324.1	22	384.5	40
TV movies/pilots	25.0	12	188.3	37	213.3	49
TV series/mini-series	127.9	19	320.3	19	448.1	38
Animation	23.8	7	23.8	3	47.6	10
Documentaries/ broadcast singles	14.5	57	0.4	3	14.9	60
Totals	251.6	113	856.8	84	1,108.5	197

Source: http://www.bcfilmcommission.com

but there was also significant foreign investment in feature films and made-for-TV movies. Over the years, many well-known and successful productions have been made in Vancouver, including cinematic releases such as *Intersection*, *Jumanji*, *Deep Rising* and *Snow Falling on Cedars*, and series including *Millennium*, *Police Academy*, *Stargate*, *The X-Files* and *Smallville*.

As a location, Vancouver has proven to be extremely well placed to benefit from increasing levels of runaway production. The city is close to Los Angeles in flying terms, has obvious cultural and linguistic affinities to the USA, is in the same time zone thereby allowing easy coordination of activities between the two centers, has a mild climate which allows all year round filming, and offers a large range of different scenic locations within one or two hours' drive of central Vancouver. Beyond these attributes, however, growth has been fuelled by some powerful economic logics. In particular, the favorable U.S. to Canadian dollar exchange rate has been at the heart of the growth process. As labor accounts for a significant proportion of the total costs of film productions, when exchange rate differentials are combined with lower absolute wage rates in Canada (on average 20–30 percent lower), considerable cost savings are generated. Overall, savings of 17–20 percent can be realized by shooting in Canada, with 60 percent of savings derived from "below the line" (technical) labor, and the rest accrued from savings on location, transportation and equipment expenses.

Estimates suggest that in reality, total cost savings of up to 26 percent on U.S. production costs may be achieved in Canada, with the extra funds being accrued from various Federal and Provincial incentives (Monitor, 1999). Since 1997, a Federal tax credit has offered rebates of up to 11 percent of spending on Canadian labor involved in a production (the level rose to 16 percent in 2003). Incentives are also available at the Provincial level. In British Columbia, the current scheme offers a tax rebate of 11 percent on Canadian labor costs, with additional incentives for productions that use certain digital technologies and shoot outside the Vancouver area. Taking these incentives together, foreign productions can achieve additional savings of well over 20 percent of Canadian labor costs through this route.

The Vancouver film industry district is constituted by a mass of small firms and individuals that coalesce together with labor provided by the various union locals on a project-by-project basis. These networks of small firms are constantly evolving and changing. However, a brief look at the definitive Vancouver film industry directory provides a feel for the scale of activity. A survey of the 1999 Reel West Digest reveals the following numbers of individuals or companies advertising their services under various categories: design and construction, 22; distribution and sales, 13; graphics and photography, 40+; motion control, 6; music, 29; post production 45+; production companies, 70+; production equipment, 50+; special effects, 26; stunt performers and coordinators 12; and studios, 17 (Reel West Productions, 1999). In addition, there are literally hundreds of other companies gaining a portion or all of their business from the film industry, listed under a general "service" category. Such services include hotels, caterers, accountants, lawyers and insurance companies.

The small size of the majority of these entities can be illustrated by the fact that in late 1997 the leading local production company employed 12 staff, the largest visual post-production house employed 150 (far more than any competitor), the largest sound post-production house employed 50, the leading equipment supplier employed 33, the two leading camera rental firms employed 15 each, the largest studio had 9 direct employees, and the leading, and by far the largest, home-grown animation company had 250 staff. While some of the larger companies may be externally owned, such as the two leading camera companies which are both Hollywood-owned, and the two largest electrical equipment companies (one Toronto-based, one Hollywood-based), by far the majority of companies are small locally owned enterprises that have grown up to meet the demands of increasing levels of location shooting. Given the large distance between Vancouver and the nearest large-scale production centers (Toronto and Los Angeles), most firms are very much locally embedded in terms of their business relationships, with many undertaking all of their business within the Vancouver metropolitan area.

That being said, the genesis and emergence of the Vancouver film industry has largely depended on the spending decisions of a small group of large U.S. studios and media conglomerates (see Figure 11.4). The studios very often have complete financial and creative control over the projects being filmed, with control being exercised indirectly through subcontracting networks rather than through direct ownership of production entities or facilities in Vancouver. In some instances, key production personnel may be sent to Vancouver to oversee or manage particular projects through a temporary subsidiary. In others, the complete production process may be subcontracted to a Vancouver production company. Whatever the chosen strategy for dispensing these funds, however, U.S. corporate capital backs an estimated 80 percent of production in Vancouver.

The small-firm networks of the Vancouver film industry thus rely ultimately on funding from U.S. studios and TV networks. The key people who engage with the financial decision-makers in Los Angeles through extra-local networks are Vancouver-based producers. Over the past twenty years, many have built strong personal relationships with Hollywood studio executives. In addition to the efforts of individual producers, Vancouver labor unions and the British Columbia Film Commission have also made concerted efforts to develop new relationships with contacts in Los Angeles. A relatively small group of key actors in Vancouver, including producers, studio managers and BCFC officials, are

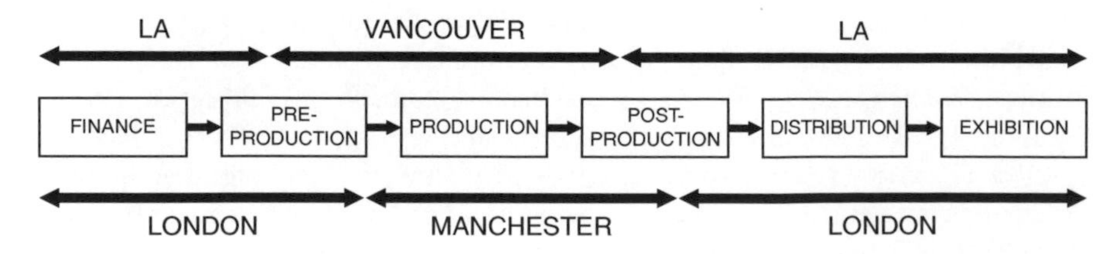

Figure 11.4 Vancouver and Manchester in the film production system

thus crucially embedded in extra-local personal relationships that facilitate the continued flow of capital from Los Angeles to Vancouver (Coe, 2000b).

So far we have presented a static characterization of the Vancouver film industry. However, the industry also exhibits two developmental trends that are seeing it start to encroach into parts of the production system usually associated with U.S. studios and media firms (Figure 11.4). Firstly, in the post-production subsector, while upwards of 75 percent of work still goes south to Hollywood, a number of Vancouver firms able to perform these functions have emerged in the 1990s. By 1999, there were 48 post-production facilities in the Province. Secondly, recent years have seen increasing levels of locally controlled or "indigenous" production. Many Vancouver-based producers who have historically serviced American business are now making sustained attempts to finance and produce their own projects. By owning the story rights and providing at least some of the funding from within Canada, the Vancouver companies can secure some of the distribution rights of the finished product, and thus a greater share of the profits. It is crucial to recognize, however, that many shows continue to require access to the huge U.S. market and funding from the studios and networks to be financially viable.

Case Study 2: linking to London? Exploring the dynamics of the Manchester film industry

Manchester has a long history of media production and distribution, beginning with the founding of the national newspaper, the *Guardian*, in 1821, and today represents the U.K.'s largest film industry agglomeration outside London and the South East. In 1998, 55.5 percent of employment in the U.K. film industry was in London, and the North West region (in which activity is concentrated in Liverpool and Manchester) contained just 5.9 percent (NOMIS, 2000).

Despite the concentration of activity in London, Manchester still occupies a significant place in U.K. film production networks. The city is home to two British television channels, BBC North and Granada, the U.K.'s largest commercial programme-maker. Its production business makes 30 percent of ITV's U.K. commissioned programme hours as well as a wide range of programmes for the BBC, Channel 4, Channel 5 and Sky. During the 1990s, government restrictions on media ownership maintained a multitude of ITV (independent network television) companies, but increased competition from satellite channels, such as BSkyB, resulted in restrictions being eased in 1993. This prompted a rash of take-overs and mergers, with Carlton, based in London, and Granada emerging as the two big players. Now, ITV is characterized by increasing concentration of broadcasting ownership, with regional companies becoming part of multinational media conglomerates (Franklin, 1997).

Changes to the regulatory framework in which these companies are operating have had a significant impact on the film industry. In response to what it saw as an "excessive degree of vertical integration" in the industry, the government introduced the 1990 Broadcasting Act which obliged terrestrial broadcasters to maintain a 25 percent independent production quota. This quota aimed to create a new era of flexibility in the production system and encourage the growth and development of independent producers, in particular those located outside London.

In the context of this new regulatory framework, Manchester's film industry would seem to have been well placed to exploit the presence of the BBC and Granada in the city and their new independent production quotas. Indeed, the early 1990s witnessed the formation of several independent production companies, post-production and facilities companies in Manchester, not least due to rationalization of workforces at both the BBC and Granada. Currently, over 400 companies employing 9,259 people are located in Manchester (CIDS, 2001). This estimate is based on more broadly defined categories of employment in film, to include firms supplying services to the industry, such as facilities companies, hence the higher figure than estimated above. A mature and highly specialized center of production exists in Manchester, with over 130 production companies, 30 post-production companies, and a wealth of firms supplying essential facilities and services.

Traditional notions of industrial districts or clusters would lead us to expect a high degree of interaction between these firms based upon evolving business and social networks. Indeed, many independent production companies and post-production facilities are spatially and socially "clustered" around the BBC, and in particular, Granada. As many of the founders of firms previously worked for either of these broadcasters, there is a degree of social interaction between actors in the industry. A superficial glance at the film industry, with a focus on production, would reveal a high number of small, specialized firms, which are flexibly and efficiently combined by producers from the BBC, Granada or an independent production company to complete a project. However, although the total production spend on television and film productions filmed in Manchester city center alone reached £25.4 million in 2001 (CIDS, 2002), the film industry "cluster" cannot be represented as a self-sufficient center of networks of cooperation and interaction. In fact, the concentration of decision-making regarding the supply of particular firms or individuals with the producer has resulted in previous working relationships and social networks driving supplier choice, rather than an open system of tendering. As a result, projects produced by Granada, or any of their associated independents, such as Red Productions, tend to use a predetermined selection of firms and freelance individuals. This situation has recently been exacerbated by the formation of 3SixtyMedia, an "independent" firm formed by the combined post-production and facilities of the BBC and Granada. Granada owns 80 percent of this company, which includes all Granada's technical infrastructure, such as studios, graphics, post-production and crews. This vertically integrated venture aims to provide a "one-stop-shop" for production, and represents a serious threat to the small specialized firms in Manchester, not least due to the expectation that producers supplying Granada with programming are expected to use this facility.

In reaction to national government initiatives focusing on the developmental possibilities of clusters of cultural activity (DTI, 2001), several regional institutions have instigated policies concerned with the media cluster in Manchester. In particular, the Film and Television Commission (NW) has supervised the development of sector-specific "networks" aimed at increasing awareness of firm's activities and facilitating cooperation and communication. All of the "networks" comprise a significant proportion of the firms operating in Manchester, but vary greatly in the extent to which cooperation between the firms has developed. The post-production "network" comprises of ten key firms, employing over 560 people with a combined turnover of £38.5m in 2001. The prime concern has been the loss of post-production work to London, even when filming has taken place in the Greater Manchester area. This process is directly related to perceived notions of the superiority of companies based in London, particularly Soho, despite many of the firms in Manchester employing highly regarded editors and receiving national and international acclaim. The firms involved in these "networks" are not particularly concerned with the development of networks within and across the groups. For them, their current networks within Manchester are sufficient for their business needs, and any false extension of these represents a waste of time and effort. Their focus is primarily on the establishment and strengthening of networks extending *beyond* Manchester. For example, the post-production group is working towards a combined campaign to market Manchester's post-production firms as greater value for money than their London counterparts, aimed at London producers.

The importance placed on extra-local networks by the film industry in Manchester demonstrates the need to take a more holistic view of this agglomeration of activity. Neglecting the significance of different forms of networks between Manchester and London reduces understanding of the role of Manchester firms in the whole production system. Several developments in the industry need to be considered in relation to these extra-local networks. Firstly, despite Granada's origins as a Manchester-based company, it has increasingly been shifting commissioning and production control to its London office. Now, very few programmes are directly controlled from Manchester. Secondly, although making claims otherwise, the BBC has not been increasing production in the regions. BBC North continues to produce a variety of niche productions, such as religious programming and regional news. These two developments have resulted in increased centralization of programming decision-making in London. As a consequence, many firms in Manchester spend considerable time and money making trips to London to maintain networks with producers and commissioning editors.

In relating this to Figure 11.1, it is clear how important the networks initially between financing and production, and finally between production and distribution, are to firms and individuals in the whole industry. The locus of power in the U.K. film industry lies in London, so for many Manchester firms their very survival depends upon their extra-local networks. In the case of terrestrial television, financing decisions are taken in London, but for satellite television and film, decisions are taken in Los Angeles by media conglomerates which are then articulated through representatives in their Soho offices.

Unlike the case of Vancouver, distances between Manchester and other centers of media production, such as Liverpool, Bristol or London, are relatively small. This greatly impacts on the ability of the film industry in the city to retain, or encourage, the use of its own facilities for filming in the area. Much production in Manchester is directly governed by producers in London, who may or may not be present on set, and is crewed by preferred London firms. Often, equipment is brought from London, or hired from national or international rental firms, despite the presence of local firms. This was clearly demonstrated by the broadcasting of the 2002 Commonwealth Games in Manchester which was produced by the BBC using very few Manchester-based firms. As the filming was governed from BBC headquarters, London firms and crews were used, despite the event essentially aiming to emphasize Manchester's importance as the U.K.'s second city.

In contradiction to expected observations of small, highly embedded firms interacting to produce televisual content, Manchester's film industry is based upon key networks connecting firms to centralized nodes of financing and distribution beyond the city (Figure 11.4). While several firms are prospering from strong networks between themselves and the broadcasters and successful independents in the city, the remainder are actively seeking to create or maintain crucial extra-local networks.

Conclusion

Our consideration of the Vancouver and Manchester film industries has highlighted the need to look beyond *production* to understand the inherent power relations underlying the whole production system. While the actual processes of production tend to involve the complex interaction of firms and individuals in specific locations, the key stages of finance and distribution upon which they ultimately depend are increasingly operating on a global scale. The film industries in Manchester and Vancouver both demonstrate the crucial importance of networks between production firms and sources of financing and distribution.

Two sets of policy implications follow from our analysis. First, although not discussed at great length in this chapter, the importance of cultural politics must be stressed. As global media conglomerates grow in size and extent over time, their power over financing and distribution will increase with serious implications for film industries across the globe. Of particular concern to national film industries is the issue of maintaining cultural diversity in the face of potential domination by U.S.-owned media conglomerates. These worries present a significant challenge to national and supra-national policy makers as they try to negotiate conflict between cultural and economic objectives. For instance, the European Commission's encouragement of "Television Without Frontiers" has led to a series of regulatory policies designed to both strengthen localized media production and to prevent the "invasion" of American culture in the form of quotas on European content.

Second, while working both against and within supranational policies related to communication and culture, there is a growing sub-national agenda focusing on the potential contribution of cultural industries to drive economic development at the local scale. The film industry in particular is often heralded as a new growth area, demonstrated by increased institutional interest and policy initiatives. Recent policy making in the U.K., for example, tends to overemphasize the relative importance of the production process, resulting in initiatives that encourage small-firm networking within cities. In many ways, this fails to recognize the crucial significance of finance and distribution, and conflicts with recent regulatory changes that have allowed increased media consolidation in the U.K.

The success of Vancouver's film industry has been attributed to more insightful policy interventions by their film commission which has actively encouraged the development of networks between Vancouver's producers and key decision makers in Hollywood. This demonstrates the crucial importance of recognizing the current and potential networks connecting localized centers of production with global media conglomerates. In addition, a greater awareness of the high-risk nature of the industry is required in order to reduce over-expectations of growth in this sector and its potential to contribute to economic development more generally. Overall, the cultural and economic dominance of Los Angeles and other global centers of power, such as London, seems only set to increase, posing significant challenges to regional centers of production.

Note

1 For the remainder of this chapter, for brevity's sake the term "film industry" will be used collectively to refer to the production of not only feature films, but any audio-visual production made for television (in all forms – network, pay-TV, cable), video or theatrical release.

References

Aksoy, A. and Robins, K. (1992) "Hollywood for the 21st century: global competition for critical mass in image markets," *Cambridge Journal of Economics*, 16: 1–22.

Barnatt, C. and Starkey, K. (1994) "The emergence of flexible networks in the U.K. television industry," *British Journal of Management*, 5: 251–60.

Basset, K., Griffiths, R. and Smith, I. (2002) "Cultural industries, cultural clusters and the city: the example of natural history film-making in Bristol," *Geoforum*, 33: 165–77.

Bathelt, H. (2002) "The re-emergence of a media industry cluster in Leipzig," *European Planning Studies*, 10: 583–611.

Blair, H. and Rainnie, A. (2000) "Flexible films?," *Media, Culture & Society*, 22: 187–204.

Browne, C. (1979) *Motion Picture Production in British Columbia, 1898–1940*. Victoria: British Columbia Provincial Museum, Heritage Record No. 6.

Christopherson, S. and Storper, M. (1986) "The city as studio; the world as back lot: the impact of vertical disintegration on the location of the motion picture industry," *Environment and Planning D: Society and Space*, 4: 305–20.

—— (1989) "The effects of flexible specialization on industrial politics and the labor market: the motion picture industry," *Industrial and Labor Relations Review*, 42: 331–347.

Coe, N. M. (2000a) "On location: American capital and the local labour market in the Vancouver film industry," *International Journal of Urban and Regional Research*, 24: 79–94.

—— (2000b) "The view from out West: embeddedness, inter-personal relations and the development of an indigenous film industry in Vancouver," *Geoforum*, 31: 391–407.

—— (2001) "A hybrid agglomeration? The development of a satellite-Marshallian industrial district in Vancouver's film industry," *Urban Studies*, 38: 1753–75.

Cultural Industries Development Service (CIDS) (2001) "Manchester's creative industries are shaping the city's future as a European regional capital," promotional document. Manchester: Manchester City Council.

—— (2002) "Introduction to Manchester"; available online: <www.cids.co.uk/mcr> (accessed 01/05/03).

Department of Trade and Industry (DTI) (2001) *Business clusters in the U.K. – A First Assessment*. London: HMSO.

Franklin, B. (1997) *Newszak and New Media*. London: Edward Arnold.

Grabher, G. (ed.) (2002) "Production in projects: economic geographies of temporary collaboration," *Regional Studies*, 36 (3), Special Issue.

Herman, E. and McChesney, R. (1997) *The Global Media: the New Missionaries of Corporate Capitalism*. London: Cassell.

Krätke, S. (2002) "Network Analysis of Production Clusters: The Potsdam/Babelsberg Film Industry as an Example," *European Planning Studies*, 10: 27–54.

Louw, E. P. (2001) *The Media and Cultural Production*. London: Sage.

Mermigas, D. (1996) "Still to come: smaller media alliances," *Electronic Media*, 11: 18.

Monitor (1999) *U.S. Runaway Film and Television Production Study Report*, commissioned by the Directors Guild of America (DGA) and Screen Actors Guild (SAG). Monitor Company, Santa Monica, California.

NOMIS (2000) "SIC coded data"; available online: <www.nomis.co.uk> (accessed 12/05/00).

Robins, K. and Cornford, J. (1994) "Local and regional broadcasting in the new media order," in Amin, A. and Thrift, N. (eds) *Globalization, Institutions and Regional Development in Europe*. Oxford: Oxford University Press.

Scott, A. J. (2000a) "French cinema: economy, policy and place in the making of a cultural-products industry," *Theory, Culture and Society*, 17: 1–37.

—— (2000b) *The Cultural Economy of Cities*. London: Sage.

Storper, M. (1989) "The transition to flexible specialization in the U.S. film industry: external economies, the division of labor and the crossing of industrial divides," *Cambridge Journal of Economics*, 13: 273–305.

Storper, M. and Christopherson, S. (1987) "Flexible specialization and regional industrial agglomerations," *Annals of the Association of American Geographers*, 77: 104–17.

Storper, M. and Harrison, B. (1991) "Flexibility, hierarchy and regional development: the changing structure of industrial production systems and their forms of governance in the 1990s," *Research Policy*, 20: 407–422.

Tempest, A., Starkey, K. and Barnatt, C. (1997) "Diversity or Divide? In Search of Flexible Specialization in the U.K. Television Industry," *Industrielle Beziehungen* 4: 38–57.

U.K. Film Council (2003) *Film in the U.K. 2002 Statistical Yearbook*; available online: <www.filmcouncil.org.uk/usr/downloads/statisticYearbook2002.pdf> (accessed 17/07/03).

Part V

Peripheral regions and global markets

12 Miniature painting, cultural economy and territorial dynamics in Rajasthan, India

Nicolas Bautès and Elodie Valette

The cultural economy of a country like India is rather different from what is often referred to as the post-fordist cultural economy in the more advanced capitalist countries (Amin, 1994; Crane, 1992). Rajasthan, one of the poorest and least industrialized states of the Indian Union, bases a major part of its urban economic development on tourism and thus cultural products make up a significant and growing contributor to this tourist economy. For several cities in Rajasthan, and among them the city of Udaipur, cultural products are not only economic assets, but more widely contribute to a process of building and rebuilding territorial identities. However, whilst a potent force for development and an important source of both income and identity the unstable and unpredictable nature of consumer demand and tourism markets mean that even small fluctuations and changes can greatly affect this fragile economic development.

The first questions in this chapter will focus on questioning the roots of the cultural economy of a city via an analysis of miniature painting production in Udaipur. Here we will be asking questions about how tourism and cultural production interrelate and how they may lead to economic development. Crucially we will also be asking whether this economic development benefits all in the local society (Urry, 2002). From the analysis of the different types of agents and of their relations, we will question the very nature of the productive system, trying to underline its local specificities. Among these, the high flexibility of the labor market seems to handle the uncertainty of the tourist demand for art works. Subsequently, the part of the painters in the local productive system seems to be difficult to assume and interesting to analyze: first because of the difficulty of making a living from painting; and second because of the necessity to change the art according to the changing tastes and demands of the tourists. Thus we will be interested in understanding how can one interpret the uncertainty, the necessary flexibility and the risks borne by artists in developing countries, and given this whether we really expect them to act as "substantive conditions of innovations and self-achievement" (Menger, 1999) or whether artistic production should be considered nothing more than a part of an economic system, where artistry only relates to its selling capacity.

The origins and development of miniature painting in northern India and Rajasthan

Miniature painting is a major Indian art form. As with other forms of Indian pictorial art, it is essentially a religious practice, traditionally dedicated to representing the principal deities of the Hindu Pantheon. These small-scale, meticulously detailed paintings began to appear in India, Bangladesh and Nepal in around the ninth century. At that time, they largely consisted of illustrations of Buddhist holy texts—depictions of the major episodes in the Buddha's life, often portraying narrative sequences from the *Jakata*. This form of painting flourished in western and northern India during the eleventh century, along with traditional mural painting. It emerged as the product of diverse influences, combining Hindu icons, Islamic symbols, and the Persian miniature tradition, thus exemplifying a process of cultural syncretism characteristic of Rajasthan, not only in painting but also in architecture, music, and literature.

The Rajput were at the center of miniature-painting production as it emerged in the major ancient royal cities of Rajasthan. Belonging to the *ksatryia* Hindu caste that comprised the Indian warrior class, the Rajput began to establish their power in the northern provinces of India in the sixth century, either by creating new kingdoms or conquering existing ones. The development of a specifically Rajput style of miniature painting can be traced to around 1600, by which time the Mughal forms and techniques had been so thoroughly assimilated as to be almost imperceptible.

Miniature painting: a sacred art under royal patronage

Until India's independence from Britain in 1947, painting was primarily considered to be a craft dedicated to the production of sacred images. In general, artistic production was a sacred activity, practiced by specialized craftsmen and sponsored by royal power. The production of miniature paintings was subject to very strict codes within the exacting vectors of religious and royal power. Indeed, the exercise of this power shaped all aspects of miniature painting, from the methods of production to the subject matter and modes of representation. This sacred context precluded any relationship between painting as an art form and painting as a commodity form.

As members of the caste of craftsmen, painters were employed by royals and by the temples. Artistic production was entirely dependent on the patronage of kings and priests. From the seventeenth to the nineteenth century, patronage by the Rajput sovereigns of Rajasthan resulted in a proliferation of different styles. For the most part, this artistic output was dedicated to poetic and religious themes, drawn from the large body of Hindu sacred literature, as well as to royal portraits, depicting sovereigns, members of the royal families, and scenes from the royal courts.

As a mediator between heaven and earth, the artist traditionally occupied an integral social position. He did not view his art as a continuous accumulation of knowledge or as a talent with which he was personally endowed. Artistic creativity and ability originated and drew its strength from the artist's relationship to the gods. The artist remained anonymous, leaving his works unsigned. According to this conception, which persists to some extent in both urban and rural areas in contemporary India, art is the expression of a religious vision of life. The artist's product represents an offering to the gods as well as the fulfillment of the human duty to the divine.

Traditional artists, therefore, did not produce miniature paintings for a commercial art market, either at a local, regional, or national level. Artistic output was not intended for sale, but rather for exchange in a royal and religious gift economy, even if some Brahmin, Rajput, and wealthy merchant families did collect miniature paintings as objects of value outside a sacred context.

From religious to commercial art: Indian independence and painter autonomy

The development of a cultural economy based on miniature paintings is a relatively recent phenomenon that emerged out of colonial administration and post-colonial nation formation. The British Raj led to the articulation of new forms of national cultural politics and the creation of public art schools, which stimulated artistic production in India and increased foreign recognition of Indian art. In 1947, national independence precipitated major upheavals in India's social, political, and economic systems, facilitating the development of new economic sectors. The cultural field was directly implicated in these changes, with Indian art becoming more and more integrated into international commercial networks.

In the particular case of Rajasthan, these changes were even more dramatic. The state was created in 1948 through the consolidation of 22 different principalities. In each of these formerly independent kingdoms, the deposition of the sovereign induced sweeping socio-economic transformations. The new dispensation had a direct impact on artistic production, introducing economic incentives for production, and modifying the social status of art and the artist. The destitution of royal power had a profound impact on the patronage system. Even though financial compensation was paid to the Rajput aristocracy by India's central government for their loss of administrative and military power, the sum was not sufficient to maintain the former levels of expenditure of the Maharanas and their families. As a consequence, royal patronage of artistic production became far less lavish than it had been.

The relationship between the royal and noble families and the craftsmen, who were financially dependent on them, is modified by the disruption of the social system. For the first time, these painters try to earn a living through artistic production. Art, once considered an esoteric craft belonging exclusively to a local or regional elite, becomes an activity structured according to the principles of Western art, which place the individual qualities of the artist at the center. From then on, art steadily became an economic activity, structured at the national level by cultural policy.

This transformation in the nature of art is part of the foundational process of the nascent Indian nation. Its evolution must be seen as part of the establishment of a modern tourist trade in Rajasthan, beginning in the 1950s. In post-independence India, modern tourism gradually supplanted the earlier forms of travel and accommodation of the British colonial elite, resulting in a rapid integration of local artistic production into larger economic networks.

Tourism and the rise of a cultural economy in Udaipur

Beginning in the 1980s, the local economy based on miniature painting in Udaipur was stimulated by the rapid increase in the number of international tourists visiting the old royal cities of Rajasthan. This global process, through which Rajasthan emerged as the primary Indian destination for international tourists, must be understood in terms of several national dynamics. For example, political and religious conflicts in Kashmir led to a strong reduction in the number of visitors to Kashmir, where the lion's share of the tourist trade had formerly been directed. In addition, during the 1990s, the Indian government implemented a number of economic reforms of considerable scope. Beginning in 1991, economic liberalization created a multitude of new opportunities and possibilities. In spite of its instability, the new economic atmosphere, with its emphasis on the globalization of trading networks, was favorable to the market in miniature paintings, which could capitalize more effectively on the taste in Western countries for the products of "exotic" cultures (Saïd, 1978). This fancy for ethnic culture is typical of the burgeoning culture of cosmopolitanism in many Western urban centers.

Tourism in Udaipur and representations of India

Tourism has become the main provider of employment and the engine of activity in Udaipur. Indeed, tourism and painting are closely interrelated: the development of an economy based on painting is dependent on the significant and steady consumer demand provided by the many tourists visiting the city. However, a reverse process is also important: the development of tourism in Udaipur is predicated on the stimulation of tourist desire through the distinctive representations of India, of Rajasthan, and of Udaipur. As A. J. Scott notes:

> Not only are there many different centers of cultural production in the modern world, but each also tends to be quite idiosyncratic in its character as a place. This idiosyncrasy resides in part in the necessary uniqueness of the history of any given place, and in part in the very functioning of the cultural economy which in numerous instances, through round after round of production, becomes ever more specialized and place-specific.
>
> (Scott, 2000: 7)

Tourist development depends on representational economies that were first established during the colonial period, and which are perpetuated and reconfigured, in post-independence India, by international tourism companies.

Initially, the image of the city of Udaipur was determined by the historical domination of Rajput power, which founded the city in 1559 and turned it into the capital city of an already powerful kingdom. Udaipur's heritage is linked to its ancient role as a capital city, which persists in contemporary constructions of local identity. However, today, this heritage has been folded into the identity of the state as a whole. The Rajput warrior, lying low in the hills and defying invaders, and the magnificent Rajput palaces, are the raw material first for a British colonial construction of India, then for a more global imaginary, with the development of international tourism. Successive visitors to Udaipur – from Moghal and British invaders to itinerant travelers, from individual Indian and foreign tourists to tourism companies – have all contributed to the construction and dissemination of these images.

Rajasthan and the Rajput clans are closely associated in local and global imaginaries. This identification is as essential to the city's image as is any contemporary reality. The city represents cultural capital that can be marketed and sold (Kearns and Philo, 1993). Among the numerous forms of this cultural currency, painting is central. Indeed, it is a well-identified sign, representing both local and Rajasthani royal culture. Painting contributes to the image of the city and its most famous sites, increasing their power to attract tourists, through aestheticized representations of the city itself, its architecture, and its natural surroundings.

The influence of tourist activity on painting production

Painting production is largely stimulated by tourism. Miniature paintings, in particular, come to embody the places visited by the tourists, and provide the tourists with a representation of the city in keeping with their expectations and imaginings. Painting is strongly linked to the production of signs, which is the foundation of tourism. It takes part in what Baudrillard calls a logic of consumption ("logique de consommation"), in which signifiers and sign-values are constrained and rationalized. There is the double constraint of signification and production: art is part of an economic system, as well as its signification (Baudrillard, 1970). As Baudrillard notes: "culture falls under the influence of the same competitive demand for signs as any other category of objects; and it is produced according to this demand." Thus:

> Culture happens to depend on the same mode of appropriation as the other messages, objects, images of everyday life [...]; that implies successions, cycles, pressure for fashion renewal, and substitutes the exclusive practice of culture considered as a symbolic system of meanings for a diverting and combinatory practice of culture seen as a system of signs.
>
> (Baudrillard, 1970: 163, *authors' own translation*)

This so-called postmodern culture does not affect its audience "through the formal properties of the aesthetic material" (Urry, 2002) but as a cultural reference. Postmodern cultural forms are not purchased in order to be contemplated, but rather as signs or representations of a city, a culture, a territory. The miniature painting functions as place-holder and cultural reference.

Subsequently, art painting changes according to the dictates of tourist consumer demand, and the corresponding ideal representation of Indian culture. Painters mobilize traditional forms of their cultural heritage to generate value-added cultural products, which are part of tourist development. In turn, the structure of demand has a modifying and regulating effect on supply. The content of the paintings remain the same – royal and sacred figures, which effectively symbolize the tourist fantasy of India and Rajasthan – thereby preserving a very important aspect of the Mewar style of miniature painting produced in Udaipur.

In a significant share of the paintings sold, subjects are selected in accordance with the expectations of Western tourists, at least as these are perceived by Rajasthani painters. Thus, a set of global signifiers, relating to India as a whole, supplements the local Rajasthani signifiers and the traditional subjects of the Mewar School. The subjects themselves are re-interpreted to match the Western imaginary construction of India. What's more, in relation to figures of local history, the more famous deities (*Sîva* or *Saraswati*, the goddess of Art and Knowledge) are over-represented. Similarly, some painters incorporate neo-hippy symbols into traditional subject matter, for instance cannabis leaves, or Sîva smoking a *shilom*.

Furthermore, the increase in demand requires an increase in supply. This extension of the supply modifies the very nature of what is sold. Miniature paintings are now available in many different sizes and at a wide range of prices, in order to appeal to as many kinds of customers as possible, from wealthy foreigners traveling on package tours to young backpackers. Also, the representation of key figures of the Indian heritage (gods, kings, Udaipur palaces, such as the City Palace or the Lake Palace) is extended from painting to derivative products: lithographs, second-rate reproductions, and even magnets. At the same time, though less visibly, a distinct contemporary art is being developed by certain painters who are re-interpreting the traditional sacred and royal motifs through a modern aesthetic.

The imperative to broaden the scope of an ever-fluctuating tourist market and the constant evolution of commercial demand lead to a degree of specialization in the production system. With the specialization of certain activities or products, the production system becomes more and more intricate, in an attempt to balance local (supply/production of paintings) and external (tourist demand) factors. These various processes, partaking of local and global logics, institute a cultural economy based on a double economic modality: tourism is a local activity that participates in a global dynamic. Tourism provides an outlet for the local production of miniature paintings, which, in turn, attract international tourists by stimulating their desire to experience an "authentic" India.

Of course, each also has an existence of its own: painting still has a cultural life independent of tourism, and tourism thrives without pictorial production. Still, their significant links demonstrate that the two partake of a single dynamic. As discussed above, cultural reference is the very object and purpose of tourist activity, as well as the basis of the system of painting production. Furthermore, miniature paintings can be seen as metonymic objects *par excellence*, which the tourist, defined here as the consumer of signs, specifically associates with the territory. In short, miniature paintings have a referential value that matters more than any intrinsic or artistic value.

Beyond the development/use of the territory as an object of tourism, the role of location is crucial (Urry, 2002; Bautès, 2003a, 2003b). The key tourist sites of the town center are the core of economic activity. One can thus construct a map of the productive city, defined as a grouping of streets, buildings and pioneer locations that are conducive to innovation (Micoud, 1991). The miniature painting production system is located in precisely such places. All the agents work and interrelate in a common physical space – where exchanges, gatherings, imitation and emulation, and competition are fostered – which contributes to the growth in the wealth and complexity of local economic activity.

Nevertheless, the sites of tourist practices cannot be strictly confined to the places where painting takes place. The painting industry is part of a larger production system that includes many other fields of economic activity. This system relies on the ever-growing number of tourists visiting the city each year (more than 100,000). In Udaipur, the hotel business, no less than the miniature painting industry, participates in the process of producing and disseminating cultural representations. These activities stimulate as well as partake of the development of the territory, the culture, and tourism. However, this cultural economy draws its specificity from the fact that it is located in a developing country, where certain distinct processes can be identified. This fact must be kept in mind, as we turn to study the various agents involved in the production and sale of miniatures.

The miniature painting production system in Udaipur

The painting industry as a whole can be understood as a complex patchwork of agents, each behaving according to their specific roles and goals. In the city, there are at least 200 painters, and almost twice as many other people involved in the industry. Among these agents, nine distinct "figures" can be identified. From an analysis of these different types of agents and their interrelations, the determinate characteristics of the productive system will emerge, and the role of the painters in this flexible labor market will be questioned.

The painting economy is based on an unstable network of agents. The nine types elaborated below are distinguished according to hazy, overlapping and constantly shifting boundaries. In this account, they are only models of the complex, transient, and flexible system of local relations. Before studying the way these nine figures interact, their specific role and importance in the whole system has to be defined. This entails an analysis of the way they work and the places in which they are established.

(1) Official representative of the local royal dynasty: a central figure in the miniature painting industry

Sri Arvind Singh of Mewar is the head of an academy, organized by Trusts, in which charity and commerce are combined. As the official descendant of the local royal dynasty, he is the central agent in Mewar-style miniature painting. Although he has never formally reigned, this royal figure is a vehicle for the material and symbolic heritage of the territory and to its most ancient traditions.

The store/workshop owned by the Maharana Mewar Heritage Trust mainly displays traditional Mewari paintings, but the offerings tend to be more diverse. Indeed, a large range of paintings is available, from very expensive ancient Mewar paintings to miniature paintings from other Rajasthani schools, and even Persian works. The store employs many people, some of them handling sales, others painting in front of the customers. The salesmen are provided with language instruction to enable them to communicate with a wide variety of foreign tourists. Wealthy Western tourists are the targeted customers because they are ready to buy high-quality works even though they know little about Indian painting. This institution, and the paintings which are produced here, serve to enhance the image of Udaipur as a city with a strong link with its past. The paintings on sale echo the traditional paintings displayed in the City Royal Palace museum.

(2) The emporium owners

The emporiums are huge stores, in which many painters are employed, often demonstrating their skills in front of customers. One could describe these places as "cultural supermarkets," targeted at tour operators who are generally brought there by official guides. Most of the time, the owner of the showroom is not a painter, or at least has stopped painting. The agents working there are mainly Rajputs, Jains or other merchant castes. They have taken advantage of the steady increase of tourism in Udaipur and the growth of demand for consumer goods. They often began as a single small shop, which specialized in painting. Since the beginning of the 1990s, some of these shops have started to transform themselves into big commercial enterprises offering a broad range of Rajasthani souvenirs: clothes, sculptures, antiques, and, most of all, paintings.

These stores were previously located within the city walls, mainly on Lake Palace Road, which was almost entirely devoted to this kind of store. However, they have been progressively relocating to new parts of the city. Today, the emporiums are large enough to accommodate the large tour groups, whose buses can be parked on the big parking lots built nearby.

Some entrepreneurs have even opened such stores in other Rajasthani tourist locations, mainly Jaipur, Jodhpur, and Jaisalmer. Rajput or Rajasthani family networks, according to very common inter-caste and inter-family forms of solidarity, provide financial assistance for the establishment of "chains" of stores, with detrimental effects on smaller businesses. In many tourist cities of Rajasthan, the larger commercial networks tend to absorb smaller stores, so as to be able to set prices and the styles of the paintings.

(3) Local sellers and national suppliers

In the old city, on some of the most popular tourist streets, some of the shops are owned by art dealers who buy paintings from passing sellers and individual painters. Inter-community links facilitate transactions, since belonging to the same caste will generally promote mutual confidence and solidarity, but they are far from being the only elements that are operative in a global market like that in miniature paintings.

(4) Gurus – traditional painters and sellers

These agents help ensure continuity in traditional miniature painting. In spite of a high degree of social heterogeneity, most of them belong to Hindu castes. Many of them work out of their own houses in the oldest parts of the city, in the residential areas of the inner city, or in a district called the *artist colony*, near the outer walls of the city. In spite of their modest way of life, they occupy the position of *guru* for students/disciples to whom their skill and knowledge is passed. They sell their own works and those of their disciples to galleries or, less frequently, directly to tourists. Their income is irregular since it is subject to the volatility of tourist demand.

(5) Street sellers

They can be found in the center of the ancient city, on City Palace Road, and clustered around places frequented by tourists. With minimal physical infrastructure, they are extremely mobile at a regional level, settling in the areas where the economic dynamics are the most favorable at any given moment. They often belong to the lower castes. A lack of financial security often prompts them to sell several types of products. The most frequent example is the sale of reproductions of modern works that are well known nationally, for example those of B. G. Sharma. Aimed at Indian tourists, these products are sold to the street sellers by suppliers from Udaipur and other northern Indian cities. Even if a few street sellers are still painters, most of them are primarily tradesmen. Paradoxically, they often have to stop selling miniature paintings because of the presence of specialized shops. Therefore, in addition to paintings, they frequently sell other crafts. But competition with the workshops is not their main problem: indeed, constant pressure from the police is the primary threat to this kind of street peddling.

(6) Intermediary agents: tourist guides, taxi and rickshaw drivers, "lapkas"

These intermediary agents are key elements of the miniature painting industry in Udaipur. Whether or not they are established in a particular place, they serve an essential linking function between producers and consumers. Indeed, the commission system is the mainstay of the economy of tourist sites like the main Rajasthani cities and Delhi. This type of relation is informal. These intermediary agents all have contacts with painters by acquaintance (friends, family). These relationships are based on the exchange of paid or bartered services. Nevertheless, they can be fully integrated into the activities of producing and selling if, as is often the case, the painter pays a few of these intermediaries to direct tourists to his workshop. These agents often have another profession: they are taxi drivers in another Rajasthani city or in Delhi, or local rickshaw or taxi drivers. Their commissions supplement their generally quite meager incomes.

This system of intermediation is of central importance for the miniature painting industry as a whole. Indeed, the production system largely depends on the ability of these intermediaries to direct tourists to the miniature painting shops or stalls. The commission system is central to Udaipur's economy, since the development of the production of miniature paintings is unthinkable without these agents.

Within this group, the youngest agents have a specific role, while only a few of them have a paid and fixed activity. Usually organized in groups of friends, they are present everywhere in Udaipur, especially around the main tourist sites but also wherever it might be possible to meet tourists. Most of them have very little education, and are of modest origins. However, their capacity to relate easily with tourists, due to their age and their command of the English language, enables them to earn a living as intermediaries. This is often their first step to setting up their own businesses. However, these agents work under conditions of constant risk, in an extremely competitive environment, where they are pitted not only against other intermediaries but also the established shopkeepers. Recently, they have come under pressure from the police, who accuse them of harassing tourists, but who also try to extort money from them.

(7) Young entrepreneurs working in cooperatives or painting academies

These agents are often directly linked to the young intermediaries mentioned above. Thanks to their mastery of tourist activity and their links to other tourist agents, they try to "go straight" by setting up well-organized businesses. These young entrepreneurs do not have much experience in the painting industry, and they often find ways to learn from their peers by following the advice of more experienced painters or entrepreneurs. At first sight, the structure does not seem strongly hierarchical. Their activity combines peddling and exhibitions in the workshop. Their partners try in turn to bring tourists to visit the workshop.

With the production process open to visitors, these workshops give the impression that they are places where the tourist can not only buy paintings and see painters working, but also truly engage with them and even learn painting techniques.

These cooperatives target wealthy tourists visiting Udaipur on package tours as well as backpackers, who usually stay for longer periods and are likely to spend time learning from the painters and eventually to buy some of their works. These businesses involve more than the simple sale of paintings; they generate diversified sources of income, from payment for lessons to the "holidays" paid for by tourists seeking an exotic adventure. In this type of business location is crucial, since the level of customer traffic depends on it. As a result, these cooperatives are almost all found in the historic center.

(8) Independent entrepreneur-painters

These independent painters usually rent studios in the streets near the city center, close to the cooperatives, the academies, and the other shops. They have an intermediate position, in the social and caste systems. They usually work alone, since they rely on several networks (family networks, for example). Most of them refuse to take part in the commission system, which, according to them, would compromise their work. Nevertheless, their strategies and practices partake of a logic of ambivalence: on the one hand, artistic aspiration and the mandate, conferred by their status, to pass down local traditions through their paintings are their predominant motives; on the other, in order to earn a living from painting, they must conform to tourist consumer demand.

These artists, usually of middle age, produce a wide variety of paintings, consistent with their sense of belonging to the contemporary art movement. Their paintings are largely determined by tourist tastes – i.e. by imaginary constructions of place, and by the main signifiers of India. If a few reproductions of famous paintings by modern artists like Dali or Munch can be purchased, original creations are still the primary product. Other painters' works are also often on offer in small quantities, to increase the range of paintings available for sale.

(9) Contemporary artists

These agents consider their practice to be more of an artistic activity than a commercial one. Nonetheless, they are integrated into the system of tourist cultural economy agents. Some of them work as art teachers in the schools or in the university of Udaipur, providing them with fixed employment and a salary, which they then often supplement by selling their paintings. They are located in or near the town center.

Their paintings, often shown in art galleries, are engaged in the project of renewing the image of India, especially for tourists. Moreover, contemporary works are considered to be alternative cultural products, integrated into the tourist supply: for example, many lithographs represent the classic themes of

Indian traditional art in a modern way. This artistic movement seems to have gained in importance in Udaipur along with the city's two emerging features – tourism and modernity. These two dynamics converge in Indian contemporary art.

There is sometimes a social or political dimension to the works of these artists. Indeed, some contemporary artists integrate elements of the pictorial traditions of tribal populations like *Bhils* or *Meenas* whose culture and art forms are staunchly defended. Several artists have recently started working with NGOs dedicated to the protection of these groups of people who are living in the rural areas, facing difficulties to adapt themselves to the fast changing economy and are officially recognized as backward groups by the Indian government.

This type of cooperation challenges the monopoly on representation of the dominant Rajput classes and helps diversify the image of Udaipur. In this way, painting can have a greater impact on the local or even national level, since it contributes to the dynamic process through which new forms of power, signifiers of place, and cultural practices are asserting themselves.

The local production system

How do these diverse agents interact? The model presented here attempts to order a complex reality and render it intelligible, based on observations of recurring processes within proliferating socio-economic relations.

Flexibility and transience of socio-economic relations

Several characteristics of this production system can be determined from an analysis of the different types of agents and their interrelations. Identifying and analyzing the cultural economy that has developed around miniature painting production yields several insights as well as complications.

This analysis places the interrelation of tourism and miniature painting production at the forefront. The uncertainty and the unpredictability of tourist demand (Urry, 2002) and of the cultural production market (Menger, 1994) require a high degree of labor-market flexibility. The production system in Udaipur actually demonstrates several different forms of flexibility: Following T. Atkinson, one can identify a numerical flexibility, as well as a functional flexibility (Atkinson, 1984). In the first type, entrepreneurs "vary the level of labor input in response to changes in the level of output" (Urry, 2002). This does not involve, as in industrialized countries, the use of part-time, temporary, or short-term contracts, since institutionalized or regulated economic relations are rare. By and large, deals and socio-economic relations are based on verbal contracts, which are often irregular and transient. Nevertheless, the specific flexibility of this system is grounded in strong regularities. Economic relations, far from being clearly defined, are more likely established on the basis of interfamilial, inter-religious, or inter-caste relations and networks. In the second type, functional flexibility is currently at work in the very generalization of multiple job strategies, as demonstrated above. Most agents seldom make a categorical choice between the three central activities of production, trade, and "touting." They often combine two of them, or even all three.

Far from being a strongly compartmentalized productive system, with a strict vertical integration of activities, these socio-economic relations are complex and diversified. There are neither clear distinctions between production and selling structures, nor a marked vertical integration within the production structure itself. Each type of agent tends to assume various strategies, adapted to the imperatives of supply and demand and to relations established with other economic agents. Nevertheless, as P.-M. Menger notes: "the flexibility requirement means that a large pool of artistic workers be available, ready to be hired as long as it is necessary and to bear the costs of oversupply of labor force and of discontinuity in the career process" (Menger, 1994). Indeed, the position of the painter is highly precarious and uncertain, and often dependent on the managers of cultural production – emporiums owners, local sellers, and so on. Flexibility and transience reflect the structure of the local social system, the local balance of power, and the relations of domination between the different groups or castes. In addition, the element of chance—indicated in the contingency of the encounters of the economic agents, retailers, producers, and tourists – is considerable.

A hierarchical system

In spite of its integration into national and global dynamics, the local production system still retains specific local features. While diversifying itself and penetrating into the world's biggest economic centers, it primarily utilizes local networks of social relations, and contributes to the perpetuation of a sense of the permanence of an ancient social hierarchy. The sedimentation of interrelated historical, social, and cultural factors in local areas generate a "work ethos based upon local religious and/or cultural attitudes" (Nadvi and Schmitz, 1998). Indeed, socio-cultural identities provide a basis for trust and reciprocity in inter-firm relations (Granovetter, 1985). This local social milieu generates "an implicit code of behavior, incorporating rules and sanctions, that regulate both social and production relations within the cluster" (Nadvi and Schmitz, 1998).

It appears that, in Udaipur, the production system tends to retain the social hierarchy established by the caste system. In Figure 12.1, three different groups of agents can be identified: high castes, low castes, and indifferent castes. The high castes are at the head of the whole production system since they buy (from a large number of agents) and sell the lion's share of the production. Many agents directly depend on them and are linked to them by informal bonds. The lower castes are primarily productive agents: the quantity of what they directly sell does not compare to that of the big players.

Economic power in Udaipur, far removed from the fictions of economic rationality, is characterized by a close relationship with the hierarchical relations

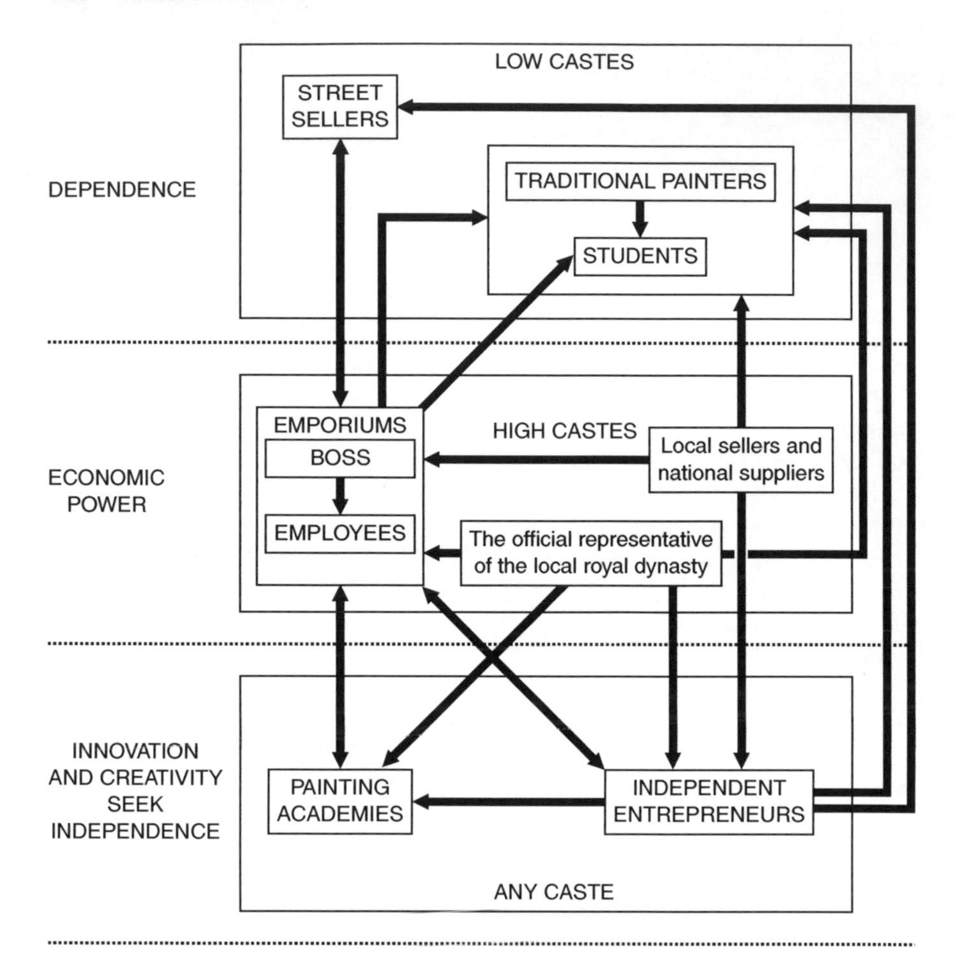

Figure 12.1 Agents and their relationships in the painting productive system

of caste. Belonging to the higher castes, the Rajputs retain most of the wealth and power in the industry. They dominate and direct painting production, while the castes of craftsmen and painters provide a flexible labor force.

Artist uncertainty and innovation

More and more painters are attracted to the city, its highly developed tourist economy, and thriving pictorial industry. The intense competition between painters tends to increase the inequalities and the biased balance of power mentioned above. Because of the constant increase in the number of painters working in Udaipur, many are excluded from the highly profitable industry.

It is clear that economic development does not provide benefits to the entire local society, and specifically to the mass of painters working in cultural production in Udaipur. In fact, one can hardly claim that "uncertainty acts as a substantive condition of innovation and self-achievement" (Menger, 1999). The painter seldom aspires to self-achievement or individual talent recognition. Only a very few painters manage to build artistic careers, based on their ability and talent alone. In order to handle the extreme uncertainty of their conditions of life, most painters, as shown in the above description of agents, try to maximize their economic independence by (a) diversifying their production in order to match the expectations of tourists; (b) diversifying their activities and assuming multiple jobs; and (c) adopting strategies of intermediation. Thus, conditions of flexibility, risk, and uncertainty can hardly be considered a context conducive to innovation, since creativity in artistic form is of little value in an economy driven by tourist demand. Nevertheless, the oversupply of artists in Udaipur can yield innovation and creativity.

A few painters, among them the group for which caste is not a determining criterion (contemporary artists, young entrepreneurs), tend to be engaged in practices of innovation. Deviating from the structures of social hierarchy linked to caste, these risk-taking independents invent complex strategies aimed at increasing the value of their activity and art; the fact that they sometimes sell their production through big selling structures (emporiums, regional sellers) is proof of the existence of a personal strategy more than a sign of their dependence. They try to diversify their production activity by combining it with direct sale of their paintings or of derivative products purchased at low cost. They also sell their production to intermediaries in order to ensure a more consistent income. Finally, thanks to long-term relationships with members of their own networks (i.e. ex-lapkas), they manage to reduce the precariousness of their situations.

This group seems to be truly capable of turning the system of local tourism and artistic production to its advantage, by implementing forms of innovation that concern either the codes and the styles of painting, or the organization of the production system itself. As creativity and innovation cannot be generated in organized work situations, the role of the individual, as a "repository of specific kinds of skills, aptitudes, and imaginative capacities" (Scott, 2000), is crucial. Indeed, local competition between agents, in the context of inadequate public cultural policy, provides the primary impetus for change in the production system.

Conclusion: a painting cluster

The painting industry in Udaipur appears at first to have certain similarities with the model of flexible production known as "the third Italy." However, the production system in Udaipur takes on a certain specificity based on its location in a developing country. Miniature painting production is certainly based on an "open community of people," but not on a "segmented population of firms" (Becattini, 1990). Vertical disintegration and flexible specialization are, in this context, merely a function of being part of informal structures. Cultural production is primarily based on unstable networks of individual or collective agents, whose interrelations are flexible, non-institutional, and transient.

Such a system can be defined as a *cluster*, a concept first suggested by Michael Porter (Porter, 1990) to describe a specific form of small-scale industrial organization in developing economies, defined by (a) a set of flexible and informal relations between economic agents; (b) a production system characterized by vertical relationships; and, (c) strong economic competition. It relies on the strength of local social, cultural, and institutional structures, operating through close and often overlapping modalities. This is of central importance in a context where "socio-cultural identity works as the base of a coded collective *savoir-faire* and of confiding relationships, all these elements stimulating arrangements and information in the production system" (Schmitz, 1990).

Although the concept of a cluster is not completely adequate to the local production system of Udaipur, there are many relevant elements in it. The role of "socio-cultural identity" is particularly significant in this context since both painting and tourism, thanks to their relationship of reciprocal stimulation, serve to reinforce the social, cultural, and architectural resources of the territory.

Acknowledgments

We would like to thank Olivier Romain for his precious help in the translation of the manuscript.

Bibliography

Amin, A. (1994) *Post-Fordism: A Reader*, Oxford: Basil Blackwell.

Atkinson, J. (1984) "Manpower strategies for flexible organizations," *Personnel Management*, August: 28–31.

Baudrillard, J. (1970) *La société de consommation*, Paris: Denoël.

—— (1968) *Le système des objets. La consommation des signes*, Paris: Gallimard.

Bautès, N. (2003) "Forms and expressions of a local heritage in a changing socio-economic context: the case of the ancient royal capital of Rajasthan with a special reference to Udaipur City," in S. Singh and V. Joshi (eds) *Institutions and Social Change in Contemporary Rajasthan*, New Delhi-Jaipur: Rawat Publications: 106–26.

—— (2004) "A city within the city. Spatial trends of tourism development in an urban context: Udaipur," in C. Henderson and M. Weisgrau (eds) *Tourism and Cultural Politics in India: The Case of Rajasthan, India's "Heritage" State*, in press.

Beach, M. (1992) *Mughal and Rajput Painting. The New Cambridge History of India; I, 3*, Cambridge, Cambridge University Press.

Becattini, G. (1990) "The Marshallian industrial district as a socio-economic notion," in F. Pyke *et al.* (eds) *Industrial Districts and Inter-Firm Cooperation in Italy*, Geneva: International Institute for Labour Studies, pp. 37–51.

Cadène, Ph. and Holmstrom, M. (eds) (1998) *Decentralized Production in India. Industrial Districts, Flexible Specialization, and Employment*, New Delhi: SAGE – French Institute of Pondichéry.

Crane, D. (1992) *The Production of Culture: Media and the Urban Arts*, Newbury Park, CA: Sage.

Granovetter, M. (1985) "Economic action and social structure: the problem of embeddedness," in *American Journal of Sociology*, 91: 481–510.

Kearns, G. and Philo, C. (eds) (1993) *Selling Places: The City as Cultural Capital, Past and Present*, Oxford: Pergamon Press.

Menger, P.-M. (1994) "Artistic concentration in Paris and its dilemmas," *Villes en Parallèles*: 20–1.

—— (1999) "Artistic labor markets and careers," *Annual Review of Sociology*, 25: 541–74.

Micoud, A. (1991) *Des Hauts-Lieux: la construction sociale de l'exemplarité*, Paris: Ed. du CNRS.

Murty, S. (ed.), (1995) *The Divine Peacock. Understanding Contemporary India*, 2nd edn, New Delhi: Indian Council for Cultural Relations.

Nadvi, K. and Schmitz, H. (1988) "Industrial clusters in less developed countries: Review of experiences and research agenda," in P. Cadène and M. Holmström (eds), *Decentralized Production in India*, New Delhi: Sage, pp. 60–138.

Piore, M. and Sabel, C. (1984) *The Second Industrial Divide: Possibilities for Prosperity*, New York: Basic Books.

Porter, M. (1990) *The Competitive Advantage of Nations*, New York: The Free Press.

Radhakrishna, V. K. (1995) *Art and Artists of Rajasthan: A Study on the Art and Artists of Mewar with Reference to Western Indian School of Painting*, New Delhi: Abhinav.

Saïd, E. (1978) *L'Orientalisme. L'Orient crée par l'Occident*, Paris: Ed. Seuil.

Schmitz, H. (1990) "Small firms and flexible specialization in developing countries," *Labour and Society*, 15, 3: 257–85.

Scott, A.J. (2000) *The Cultural Economy of Cities: Essays on the Geography of Image-Producing Industries*, London: Sage.

Urry, J. (2002) *The Tourist Gaze*, 2nd edn, London: Sage.

13 Profiting from creativity?

The music industry in Stockholm, Sweden and Kingston, Jamaica

Dominic Power and Daniel Hallencreutz

Introduction

The popular music industry has in the last 50 years grown to become an important global industry and a major area of economic activity; and one that operates on a variety of intersecting geographical scales. The music industry is, most often, a highly localized cultural product industry that draws on local creative milieus and cultural forms and one that has a tendency to agglomerate in urban areas (Hesmondhalgh, 2002; Scott, 1999a, 1999b, 2000). However, the business of producing, selling and consuming music is globally dominated by an increasingly concentrated manufacturing and distribution system dominated by a shrinking number of global media "majors" (Alexander, 1996; Burnett, 1996; Choi and Hilton, 1995; Malm and Wallis, 1992; Sadler, 1997; Shapiro *et al.*, 1992; Wallis and Malm, 1984). Whilst many places in the world are home to dynamically creative cultural and musical milieus, it is interesting to note that in the history of popular music relatively few have fostered groups or "scenes" that have gone on to international or global success, whether this be measured in terms of profits or cultural influence, though in this chapter we focus on financial success/gain. It is argued here that in the music industry it is not only the quality of the creative milieus that counts towards commercial success but also the links between the local production system and international circuits of capital, distribution and effective property rights.

This argument is tested using the cases of two of the world's most dynamic and creative musical agglomerations: Stockholm, Sweden and Kingston, Jamaica. Stockholm is the center of the highly centralized or agglomerated Swedish music-production system (Forss, 1999; Hallencreutz, 2002; Power, 2002) which after the U.S. and U.K. is the largest net exporter of popular-music products (Burnett, 1997) having produced a remarkable number of internationally successful musical acts – including amongst others Abba, Ace of Base, the Cardigans, Neneh Cherry, Europe, Stina Nordenstam, and Roxette. Just as in Sweden, Jamaica's music-production system is almost entirely dominated by the capital city. Fueled by the worldwide popularity of ska and reggae music, from the 1970s onwards Kingston has been the creative center for some of the world's most important and commercially successful popular music-recording artists – including Bob Marley and the Wailers, Lee Scratch Perry, Desmond Dekker and the Aces, Jimmy Cliff, Gregory Isaacs, Peter Tosh, Shabba Ranks, and Shaggy. Whilst not comparable in terms of socio-economic conditions, and of course climate, both cities are directly comparable when taken seriously as world-class centers for the production of popular-music products. The choice of the two cases for comparison then rests upon the recognition of both cases' leading edge position in the production of musical products. This choice is further validated, we believe, by the fact that this assessment rests to a large extent on the international and export competitiveness of their respective products: in short they are strong competitors in the same global market. The fact that both of the music production milieus treated here are embedded in relatively small economies, albeit radically different ones, makes the international competitiveness and export of their products more important than it may be for artistes and firms in larger economies, where domestic demand is large enough to support a profitable and fully functioning industry that need not necessarily look outside its borders in order to be viable. (One may think here of countries such as Brazil, India or China that have highly productive and profitable music industries despite the fact that their products are not significantly exported to or consumed in international markets.) These considerations, we believe, make the two cases more amenable and interesting for comparison than they may first appear.

What is particularly interesting, however, when comparing the two is that although Kingston's products have a far higher global commercial value than those from Stockholm, it is Stockholm's local production system and urban economy that makes the bigger profit in real terms. The United Nations Commission on Trade and Development calculated the recorded musical product (not including live performance and related merchandise) of Jamaica, in the main reggae music, to have had a worldwide (excluding the domestic Jamaican market – worth less than U.S.$2 million in *retail* sales in 1994; U.S.$2.2 million in 2000: IFPI, 2001a) *wholesale* value of, at least, around U.S.$1.2 billion in 1994 (Kozul-Wright and Stanbury, 1998; see also Watson, 1995; Bourne and Allgrove, 1995; PIOJ, 1997) whilst exports of recorded music from Jamaica itself totaled a relatively minuscule U.S.$291,000 (Bourne and Allgrove, 1995). Whilst the total worldwide wholesale value should not be treated as exports from Jamaica due to the fact that many successful Jamaican artistes record and produce most of their work in North America and Europe, it is clear that there exists a very large imbalance between what is happening and what could happen, and that even a relatively small increase in the Jamaican export of music would greatly add an economy whose total value of exports in 1995 was U.S.$1.4 billion. In stark contrast the Swedish case is one where export and domestic market success have led to significant benefits to the national music industry and its Stockholm heartland; Swedish net exports of recorded music in 1997 (the year for which exists the only comparable data to the Jamaican figures above) were around U.S.$411 million (SEK3,368 million) (see also Forss, 1999, 2001). Whilst these are quite rough measures of value what they illustrate is that despite a proven market of at least U.S.$1.2 billion for "Jamaican" music outside the country, only a tiny percentage trickles back to the "producing" country.

Bearing the above in mind, this chapter explores the unique bases upon which each center's strong competitive position rests and the dynamics by which this competitiveness is translated into vastly differential rewards for the respective local systems and economies. Underlying the chapter is the argument that analyses of the benefits of agglomeration, clustering, and localized conditions, etc. must also take into account the wider mechanisms by which individuals and firms profit from creativity. Few studies have attempted to study the returns to creativity and innovation arising from globally competitive cultural products centers, with far too many studies concentrating on the genesis of creativity rather than the dynamics of the large value chain. Indeed, one might say that in general studies of the competitiveness of bounded or agglomerated production systems often focus primarily on the dynamics and mechanisms of developing creativity and innovation and pay relatively scant attention to the reasons behind differential rates of return accruing to areas with similarly competitive capacities. Furthermore, few studies tackle such issues in a comparative context; not least between centers embedded in "advanced" and "developing" economies respectively.

The chapter draws especial attention to the crucial roles of intellectual property rights (IPR) regimes and attendant industrial and firm structure for the "profitability" of music. Underlying the whole article is the recognition that copyrights and IPRs are the constitutive institutions upon which the music industry's present configurations rest (cf. Dolfsma, 2000).

In recent years a series of technological innovations – such as internet-based file sharing, compressed file formats (e.g. MP3), and CD burners – have threatened the effective protection and collection of copyright revenues from music (Dolfsma, 2000; IFPI, 2001b; Kasaras, 2002; Leyshon, 2001; Pfahl, 2001). To an extent the case of Kingston presented below can be seen as a model of a "post-MP3 musical economy" where the lack of an enforceable or effective IPR regime configures the music industry and economy in a manner antithetical to the large-firm-dominated model that is common in much of the rest of the world. As we shall see, the absence of an institutional structure to protect copyright and collect royalties, etc. has meant that Jamaican "investments" and "product development" have not been protected. This has resulted in actors and firms being forced to turn over their cultural and financial capital as quickly as possible resulting in a fragmented industrial structure characterized by under-investment (financially at least) and cut-throat competition. Stockholm by contrast can be seen as an example of a musical economy centrally located in the present capitalist IPR regime: an economy where musical products are defended by a strong local regime that allows long-term investments in creative acts to be returned. The two cases then serve to illustrate that the present political economy of place where different positions within the global economy confer vastly different rewards; and that differences in, and the erosion of, the efficiency of IPR regimes can have dramatic effects on industrial and profit structures for cultural products industries.

Thus the chapter concludes by reaffirming the often noted positive correlation between agglomeration and economic and creative innovation but adds that rates of return to agglomerated economic activity and innovation vary considerably due to a variety of factors. It is concluded that in these two cases differential levels of profit repatriation can be explained by two main factors: copyright regime and administration on the one hand, and industrial and firm structure on the other.

Stockholm

Despite talk by many music journalists of the existence of a distinctive Scandinavian sound, exemplified by the likes of the Cardigans and Stina Nordenstam, the strength of the Swedish music scene seems to lie in producing Anglo-American music that is often better than the "real thing." By singing in English and fitting into well-established rock, pop, and dance genres, many Swedish artistes have produced products easily palatable to international markets and enjoyed considerable commercial success. In common with many sectors of the country's economy, the Swedish music industry is highly centered in the region of 1.5 million people that make up the Swedish capital and in specific the central city area of *Stockholm stad* (Forss, 1999; Hallencreutz, 2002; Johansson, 2000, Power, 2002.)

Though Stockholm is by no means the most predominant source of musical talent in Sweden – as many of the most famous bands originate from other cities, e.g. the Cardigans from Jönköping, The Hives from Fagersta, and Ace of Base from Göteborg – it is the place where most artistes and industry workers pursue their professional careers or progress to the next level (cf. Denisoff and Bridges, 1982). According to the most recent estimates available, there are at least 3000 professional musicians, composers, and producers in Sweden, mainly operating as freelancers (Forss 1999). Along with a far higher number of amateur musicians, these professionals constitute the basis for the specialty inputs – in the form of music creation – to the music industry. These "creators" tend to "end up" in the Stockholm region with smaller but significant numbers in Göteborg and Malmö (e.g. the three major urban areas). It is interesting to note though that a great many of these were not born in the Stockholm region. Thus, Stockholm's industry is heavily relying on "competence-input" from other regions in Sweden.

These "creators" both attract firms to, and are drawn in by, Stockholm's position as the center of the industrial system of music production and sales in Sweden. The region has a large number of both local and international music companies: around 200 record companies and approximately 70 music-publishing companies; around 50 percent of the national total. The most immediately apparent aspect of the firm structure of the city is the extent of its internationalization; or its exposure to the global majors. All the "majors" in the global recorded music industry are active in Sweden through fully owned subsidiaries headquartered in Stockholm. Since the early 1980s these corporations have strengthened their position by acquiring most of the large independent companies (e.g. Metronome, Elektra, Sonet, and Polar). When acquiring these companies, the corporations also incorporated their licensing deals and facilities for distribution, as well as affiliated publishing companies. The ten largest record companies (by turnover) in Sweden are now all owned by foreign majors and all headquartered within walking distance of one another in central Stockholm. Thus the global majors have to a large extent acquired all the central elements of Stockholm's, and Sweden's, record production and music-publishing system.

Parallel to this concentration of ownership has been, since the early 1990s, the growth of small, independent, flexible full-service publishing and production companies in the Sweden, most notably in the Stockholm region (e.g. Cheiron, Murlyn Music, Sprinkler, Maratone). A key example amongst these was the firm Cheiron which quickly became a world-renowned center for the writing and producing of pop and rock music. The firm has now broken up (after the untimely death of one of its founding members) but was in essence a collective group of producers who operated a studio in Stockholm aimed at producing global hits: the surviving members have remained very active in the industry. The group was remarkably successful in producing both successful Swedish acts (such as Ace of Base, Robyn, and Papa Dee) and international artists (such as the Backstreet Boys, Britney Spears, Bon Jovi, Celine Dion, and N'Sync). There is little doubt that Cheiron and others have been powerful role models to the city's music industry in general and have created widespread awareness of the success to be had from business areas other than the fostering of new artists most people think of as the core element of a successful music industry. Equally it is only one example of a large number of firms that have functioned as incubation units for new firms and spin-offs. During the firm's lifetime, and in particular after its break-up, numerous new firms were formed that built on expertise and contact networks built up by onetime members of the firm. Evidence shows that throughout the 1990s and today continuous new firm formation has been a feature of the city's scene and that there has been significant growth in the number of independent record companies, professional musicians, publishing companies, production companies, and music-related IT companies (see Power, 2003). Music-video production firms and music-oriented multimedia firms in Stockholm are likewise examples of associated services that are supportive of both indigenous musical product and employment, and export-driven music services provision; for example, Åkerlund & Pettersson Filmproduktion and Bo Johan Renck who have received international recognition for their videos for the likes of Metallica, U2, Iggy Pop, and Madonna. An interesting feature of this music services environment is not only that these production companies have produced songs, multimedia content, and videos, etc. for international artists but also that international artists have chosen Stockholm as their recording and creative locale. Forss (1999: 107) estimated that sales of music services to foreign clients accounted for just over 12 percent of Swedish music exports. This has, according to local artistes, strengthened the creative "buzz" in the city and given many artistes further reasons for staying in the city rather than leaving for global centers such as London or New York. At the same time, artists interviewed suggested that studio time is becoming increasingly expensive, thus having negative effects on newcomers' ability to create high quality "demos." The existence of an open, creative milieu and the growth of specialized service firms fits well with the recorded music industry's practice of subcontracting out, often internationally, as many functions as possible.

It would be wrong to say, however, that the smaller Swedish firms are in constant conflict with the foreign majors and their branches/affiliates. Rather, it appears that many synergies exist between the different types of firms which are often cemented through the numerous examples of formal "buyer–supplier" linkages found; particularly different types of project-based, socially embedded linkages. Furthermore, there exists a number of shared business-related local institutions and evidence of informal co-operative competition, such as joint projects at international fairs (e.g. Midem and Popkomm) and lobbying for standards in copyright law. In general, research suggests that the success of the independent production and post-production firms in Stockholm and the degree of formal and project-based linkages between majors and minors has had several specific effects: it has greatly enhanced domestic artists' and firms' exposure and access to international best practice; has led to increased investment in the quality and capacity of firms' technical infrastructure and know-how; and solidified associations between "Stockholm" and "quality product" in the minds and operations of European and global industry actors and firms.

An important consideration for understanding the general evolution of Sweden's commercial music industry is that a very supportive consumer base exists: domestic demand is very high with a domestic retail market worth around U.S.$322.9 million in 2000 (IFPI svenska gruppen – www.ifpi.se) and with Swedes ranking as the sixth highest per capita consumers of recorded music (IFPI, 2001a). A sizeable proportion of sales are of Swedish artistes (27 percent of sales in 1998: IFPI svenska gruppen – www.ifpi.se), suggesting both awareness of local producers and indicating a high level of quality in the domestic product. This is backed up by industry sources that stated that the Swedish consumers were considered to be highly "sophisticated" (see Porter, 1990), having in the past proved themselves to be very early adaptors to changes in musical styles and trends as well as technologies. Sources in Stockholm-based branches of the majors said that the Swedish market was used as a testing ground for new products and groups, and that success in the market was often considered an important step in the development of the firms' wider European-level sales strategies. Furthermore, Swedes have long been amongst the first users of new technologies (e.g. Sweden was one of the first countries to adopt the CD format as the standard) and have some of the highest levels of internet penetration (Eurostat, 2001) and commercial ICT usage (OECD, 2000) in the world, which has meant that the new technological possibilities shaping music distribution (and piracy) have gained rapid ground. Added to this Stockholm youth's high levels of disposable income support a highly profitable, competitive, and dynamic music and club (both nightclubs and fan clubs) scene that interviews revealed to be important to the showcasing of local, and national, talent. A&R (Artists and Repertoire – the traditional industry term for "talent scouting") executives spoken to said that the strength and diversity of the relatively compact local scene greatly helped in spotting talent; though one admitted that the A&R world's concentration on central Stockholm may be detrimental to the chances of artists who do not perform there.

Servicing this demand is a large number of retailers, most of which are parts of larger chains or subsections of extensive retail operations such as department-store chains. These retailers most usually source their product through a relatively centralized national distribution system that is located in the Stockholm region.

The city's industry is further supported by a high level of technological ability and infrastructure. The existence of the latest facilities for the completion of a project from creation to saleable product has been an important factor in keeping the largest possible part of musical products' value chain and profits in the country and city. As shall be seen in the case of Kingston, it is quite common for smaller centers' and countries' musical exports to consist of only the core idea or track which is then sent abroad for post-production, pressing, marketing, distribution, and so on. Since the largest majority of the profits are to be gained at these later stages it is important for a center to have the capacity to do as much as possible "in house." Stockholm as a city has one of the most advanced media technology and telecommunications infrastructures in the world and can handle all aspects of the production chain. For instance, despite being a small domestic market, Sweden exhibits a high density of CD-pressing plants compared to other countries (Forss, 1999: 99 estimated the export value of physical recorded media "pressed" in Sweden – CDs, tapes, records, etc. – to have been worth SEK1,483 million out of total musical exports of SEK3,368 million). The majority of these plants are located in the greater Stockholm area along with various producers of CD-pressing technology, musical instruments, studio equipment, and so on. Being close to or part of an advanced technology environment has also fostered the growth of a relatively large number of companies on the interface between the music industry and internet content and distribution companies.

Perhaps the most important factor in securing the "repatriation" of profits in an export-oriented cultural products industry is the existence of an effective intellectual property protection system. Without such a system the profits of creativity have a tendency to disappear very quickly. By virtue of a long history of artisan organizations and unions the music industry is well tended for by representative organizations and royalty-collection agencies. The domestic presence and international oversight and enforcement mechanisms of many of these organizations, combined with the Swedish state's active international protection of Swedish intellectual property, are important in minimizing as much as possible the loss of export revenue within the chaotic maze of international music licensing and consumption and, to a lesser extent, piracy. Furthermore, the domination of the majors in the export of Swedish musical products and services has meant that Swedish products have enjoyed the majors' extensive powers to globally protect IPR: the majors are the only actors with the resources and manpower to protect (or exploit) copyright worldwide.

All in all, Stockholm seems to be a clear-cut case of industrial competitiveness often developing in a clustered or agglomerated manner; that is, it takes a large number of firms that are both competing and cooperating with each other to trigger growth. The industrial and organizational structure of Stockholm's music industry is at once both highly competitive and highly cooperative, and characterized by a diversity of actors and firms with a relatively high turnover. The pace of structural, organizational, and technical change in the city's industry has been high, and it would seem as though these changes have benefitted the exports of Swedish musical products. For these authors, however, the feature of the Stockholm scene most supportive of export performance are the dense networks of inter-organizational linkages, voluntary associations, and service organizations that both secure relatively open export and distribution channels and copyright (and thus indirectly revenues back to the creative milieu). The music business, or rather the actors – be they firms or individual artists, composers, music teachers, publishers, or any other professional group – appear to have a high capacity for self-organization; that is, to create mechanisms for the coordination and promotion of joint interests, exchange information, and collective action. The Stockholm music "cluster" could be defined as a "world of production" (Storper and Salais, 1997), embedded in the wider global music industry and characterized by a complex mixture of cooperation and competition. The quality and manner of intra- and inter-firm linkages and relationships can be said to have added to the success of both Swedish products in wider markets and also the protection of export earnings involved in Stockholm's role as a major international production and music-related services center. The perhaps surprising conclusion – given the current strength of anti-globalization rhetoric – is that it seems higher or increasing levels of integration into the wider global music industry (and its concentrated power structures) have in fact added to the competitiveness and crucially the profitability of the local production system of the city's music industry.

Kingston

Kingston currently performs the function of being something of a "neo-Marshallian node" (Amin and Thrift, 1992) in the global music industry. It is at once the production and innovation center for a small but complex and dynamic domestic music market as well as being a strong global supplier of both musical product and innovation (from musical techniques such as those used in dub to entire stylistic genres such as ska, reggae, dancehall, and ragga: see Berg, 2001; Chang and Chen, 1998; Davis and Simon, 1992; Foster, 1999; Harris,

1984; Jahn and Weber, 1998; Potash, 1997; Saakana, 1980). From its initial commercialization in the early 1960s onwards, Jamaican music has evolved to have both a strong market presence and a relatively cohesive identity. Right from the start Kingston has been the creative heart of the entire country. From the marginalized west-end ghettos of the city, such as Trench Town and Jones Town (see Chang and Chen, 1998), musical styles such as ska and reggae broke forth; as well as their principal proponents such as Bob Marley, who in recent years has gained what amounts to iconic status in Western popular culture (Hussey and Whitney, 1999; Stephens, 1998) and the commercial possibilities that brings with it.

The most interesting feature from a comparative perspective of Kingston as a music production center is that, unlike almost all other centers in the world, the Kingston case is one where the global majors have almost no direct presence or role. The organizational and firm structure of the city's industry is largely fragmented and overwhelmingly dominated by small-scale, A&R, and original creation-driven firms. Although there are currently no representatives of the global majors in Jamaica, attempts have been made in the past by the modern corporations' predecessors to set up in Kingston. The first such attempt was made by CBS (now part of Sony Music) in 1976 who wanted a Kingston base from which to supply the then rapidly expanding international reggae market. Local record firms and industry actors, however, successfully blocked the move by appealing to the then Prime Minister Michael Manley's strongly nationalistic and isolationist prejudices. Manley's departure in the violent elections of 1980 and the arrival of the more liberal Edward Seaga (who had at one point set up a record company himself) into office prompted CBS to make another attempt. This time they attempted to buy one of the larger local firms, Dynamic Sounds, with the aim of turning it into a mini TNC covering the entire Caribbean area. However, the deal was eventually stopped by strict laws against TNC holdings in domestic firms.

CBS's attempts seemed to have acted as a warning to other foreign corporations and none have since tried to establish themselves on the island. The global majors have shied away from direct involvement, preferring instead to "cherry pick" already developed products. This is widely acknowledged by industry actors interviewed in Kingston to have in the past been a rather dubious, and often exploitative or even criminal, process with the majors buying complete master tapes and exclusive rights off whoever could get their hands on them.

Links between foreign firms and local firms and artists have, and are, put under further strain by cultural differences that have led to an extreme unwillingness on the part of the majors to have dealings with local firms. Industry sources suggested that many local actors have an aggressive "get rich quick" approach (see also Kozul-Wright and Stanbury 1998) that makes them unwilling to enter the sort of long-term legal and financial arrangements the big corporate players demand: in markets such as the U.S. and Europe it takes a long time and large capital investments to launch, warehouse, and distribute products, and added to this the majors tend to release products over a long period to avoid over-exposure of the type that may bore consumers. Stories abound in the majors also of the extreme difficulties of working with the often temperamental natures of those Kingston artists and firms (though one would imagine international record companies should be well used to most types of "unconventional" and "unruly" behavior by now).

Despite the lack of direct involvement by the majors it is important to note that foreign firms have in many respects been the most important actors in the history of Jamaican music and its international success. Probably the most important lead firm in the development of Kingston's music industry has been Island Records (bought by Polygram in 1998 and now the Island Def Jam Music Group, a subdivision of Vivendi Universal Group), originally set up and owned by the English-born Chris Blackwell.

The dynamism of Jamaican musical output then has largely been a story of endogenous growth in innovative activity. Exact figures on the number of firms and individuals involved in the production of music in Kingston are hard to come by due to the lack of precise figures on the industry and the informal (often illegal) nature of many firms and individuals involved. Furthermore, the tendency for actors to work through multiple labels and firms and simultaneously take up multiple roles (e.g. producers who are also artistes, sound engineers and label owners) makes it virtually impossible to disaggregate the components of the industry in a comparable fashion to that done above with the Stockholm case (where divisions of labor are much more rigid). What is certain, however, when one reviews histories of popular music, global recorded music sales, and current international and Jamaican charts is that at least in terms of professional artistes a large number – around 2,000 acts (either single artistes or groups) according to local estimates – currently reside or have their origins in Kingston.

These acts are supported by a large number of independent, most often very small-scale, record companies and labels: such as Tuff Gong, Dynamics, Penthouse Records, Sonic Sounds, and Scorcher Music. A major feature of the system is that it is relatively unique in organizational form and differs from places like Stockholm where record companies, characterized by high degrees of internal transactions, are the central production units. Kingston is characterized by a complex mix of producers (in the music industry sense of the term), record companies, and record labels with high degrees of external transactions.

The complex and fragmented production system with its multiplicity of actors, most often working on an essentially freelance basis, can be seen to have strengthened the role of interpersonal and informal contacts in the innovation process. In order to survive in the city's industry, firms and individuals realize that constant social and business interaction is important. In contrast to the trust and cooperation that are often pointed to as central to the creative milieus of small-firm clusters (cf. Scott, 2000), in Kingston this interaction is predominantly competitive rather cooperative. The diversity of small-scale firms who do not have the capabilities to fully develop products in-house means that the industry is characterized by high levels of external transactions which further embed actors in the system. External transactions and dense social interaction mean that information flows more fluidly than in many other music-production centers. This means that it is extremely hard to keep products under wraps before release and that new musical ideas and techniques rapidly diffuse (one day is not an uncommon time for new "riddims" to be widely copied by other artists and made available to the market). Thus agglomeration (of musical activities) in the urban area fosters intensified transactions and interactions that are crucial to commercially successful creative and innovation product development. Furthermore, the speed of transactions and information exchange have a definite effect on the way in which products are released. Whilst in European and American markets product release tends to be spaced out over time, in Kingston record companies pursue a policy of flooding the market in an attempt to sell as many records as possible in the shortest possible time.

There may of course be considerable benefits accruing to musical innovation in an area with a fragmented firm structure. Some writers have noted that fragmentation and lack of concentration in the corporate organization of recorded-music production may favor diversity in product type and encourage innovative processes (Alexander, 1996; Burnett, 1990, 1992, 1993; Christianen 1995; Lopes, 1992; Peterson and Berger, 1975, 1996). The level of fragmentation and often disorder in the structure and organization also has important negative consequences though. One consequence is that the lack of links to international players and their distribution channels and resources has significantly underlined the lack of capital and financing sources for upgrading and investment in new products; successful launch of new records on an international level involves a substantial investment in stock, marketing, advance shipping costs, etc. that most Jamaican firms cannot finance. Interviews revealed that those studios that cannot afford to upgrade facilities to a decent modern level have moved away from serving the "serious" music community and shifted towards a focus on recording commercial jingles and the like for radio and TV advertising. This has further reduced the production capacity of the industry and has channeled production and mixing expertise away from the quality-music products upon which the industry's long-term future rests. Firms' lack of resources to invest in advanced production and recording techniques and, crucially, information on market-demand conditions outside the country means that they rely on obsolete recording and production equipment which makes international standard digital quality and format production impossible – nearly all Jamaican releases are on tape or vinyl – which in turn closes off direct access to high-spending but quality conscious foreign markets such as the U.S. and Japan. The lack of a capacity to supply modern digital-quality products is a major problem in an international context; especially so when it is remembered that internationally CDs account for around 75 percent of the overall market and almost 90 percent of the singles market (IFPI, 2001a). The problem is not just one of funding, as fragmentation and lack of firm level exposure to the wider global industry further distances Kingston's producers from foreign markets and the modern techniques they demand. Furthermore, the lack of high-standard equipment and up-to-date techniques means that domestic post-production, remixing, video production, and related services tend to get exported. It must be remembered that it is often in these later stages of musical products' journey to the market that added value is created and profits made and that as we saw in the case of Stockholm these type of services can draw in business from outside the country that is both profitable and beneficial to the dynamics of the local center.

Thus firms must rely on the low-tech and impoverished domestic market. Demand in the domestic market for the CD format is extremely low – probably around 10 percent of unit sales (see IFPI, 2001a) – and sales of CDs almost entirely relate to foreign product. Insufficient local demand for CDs and digital quality and the market's relative poverty further hinders attitudes and abilities to upgrade: in 1999 the Jamaican music industry on the domestic market generated U.S.$5.4 million, which leaves only seven countries that generate less (IFPI, 2000). Interviews revealed that the average "hit" tends to sell between two and ten thousand vinyl copies normally priced at around U.S.$2 per unit. Once the profits are shared out over the entire value chain there remain only very small margins for those involved. This is especially true of the original producers who often see little use in spending time making recorded music due to the poor domestic incentive structure. Interviews showed that the lack of sufficient incentives to record music means that today's artists – who tend to have a "get rich quick" attitude to the business – concentrate their efforts on live performance (a live performance brings in, depending on their status, between U.S.$2,000 and U.S.$15,000 for an artist). This further focusses the majority of the industry on the domestic market and accentuates the cycle of lack of investment and ability to produce product exportable on competitive terms.

Although the domestic market may in terms of cash be relatively impoverished, it is one that is dominated and driven by highly "sophisticated consumers" who are crucial to the innovation and creative dynamism of the industry. Whilst many things go very slowly in Jamaica, others go very quickly and this can be seen to be especially true of music trends and fashions in Kingston. Successful sounds and "riddims" are copied almost instantly and it is not untypical for a sound or stylistic innovation to be totally over-exploited and out-of-date within a few weeks. The speed of production and release coupled with the speed at which consumer fashions change puts an enormous pressure on even the most established artistes to constantly come up with new sounds to satisfy audiences. The intense competition amongst the many highly flexible and adaptable distributors and retailers (many often act as both) ensures that product gets to market, often direct from the studio, at breakneck speeds. On the other hand, fragmentation makes marketing and distribution economies of scale hard to achieve, limits reinvestment opportunities, and makes it much easier for pirates and gangsters to be "involved" in the value chain.

The figures on the value of recorded-music exports from the country quoted in the introduction and indicators such as the level of investment and upgrading underline that, whilst Jamaica has the demonstrated potential to give rise to musical products with a large global commercial appeal, the local production system benefits very little from this competitiveness. The export figure of U.S.$291,000 for 1994 of course underrepresents the earnings from music products that reach the island – from the limited amount of royalties that do make it back; the "invisible" exports of pirate exporters; personal monies brought home by those stars that visit home on a regular basis. Many also consider musical success to have had a direct effect on the success of the island's tourist industry. Nonetheless, it is obvious when compared to the likes of Stockholm that Kingston is reaping extremely meager benefits from its globally competitive position in music.

The problem of small returns on high levels of musical creativity to the country's music industry is widely recognized as being in large part due to the lack of an adequate intellectual property protection regime in the country and the lack of effective enforcement of existing legislation and international protocols (cf. Bourne and Allgrove, 1995; Kozul-Wright and Stanbury, 1998). On an international level the music industry in general has serious problems in enforcing statutory claims to usage and remuneration from individually held intellectual property rights (IFPI, 2001b; Wallis *et al.*, 1999). Recent agreements under the North American Free Trade Agreement (NAFTA), the General Agreement on Trades and Tariffs (GATT), and the European Union have further extended the scope and force of copyright protection and therein further favored the global majors' "ownership" of revenues from the majority of artists' output and existing copyrighted material (cf. Galperin, 1999; Taylor and Towse, 1998; Wallis *et al.*, 1999). In most countries an array of well-organized collecting societies and industry organizations oversee copyright enforcement and collect and distribute revenues. But in Jamaica the fragmented firm structure, lack of direct links with the majors, and the underdeveloped nature of performers' organizations and industry representatives have made it especially hard to internationally enforce existing copyrights and collect revenues. There is a lack of effective industry-wide supporting/supporting institutions, though organizations do exist – such as the *Jamaican Federation of Musicians and Affiliated Artistes* and the *Jamaican Association of Composers, Authors and Publishers* – and are working to improve. Coupled with the fragmentation of the industry's firm structure this has meant the industry has little coherent voice in the island's policy or legal circles. This lack of voice and the establishment's traditional fear of, and often hostility towards, some of the industry's actors in large part explains why it is that the Jamaican government has only recently begun to officially recognize the industry's economic significance and potential (OPM, 1996) and attempt to develop a more effective copyright environment. An example of this is that it was not until 1993 that the island's copyright law was updated to an international standard. Although Jamaica has legislation and is now a member of the World Intellectual Property Organization (WIPO), protection is limited by the fact that an effective national collection agency does not exist: it is through reciprocal arrangements with other national collection agencies that rights are most effectively enforced and royalties collected. An important part of the intellectual property environment in Kingston is industry specific, and culturally related to the underground nature of musical creation, propagation, and commercial adaptation.

As Paul Gilroy (1987: 164) notes, the focus of modern Jamaican dancehall culture is not the cult of personality that drives the commodification and value chains of popular music but rather the blending and creative appropriation of others' music. Interviews further revealed that this lax attitude to musical ownership and property rights, by the standards of copyrighters, was not only a part of the current general cultural norms that operate in the Kingston scene but had also often been part of the religious beliefs that many of the Rastafarian predecessors to dancehall and the like thought of as central to musical expression. Indeed, the problems many artistes and firms have had with copyrights (particularly with pre-1980s recordings/compositions) can be related to Rastafarian anti-property attitudes. The most obvious sign perhaps of this anti-property attitude is the infrequency with which reggae stars have, or have left, wills and well-managed intellectual property portfolios; most notably the case of Bob Marley, who failed to leave a will. This sort of view of intellectual property is far removed from the view that prevails in the global industry. Lack of an effective system for collecting and protecting rights revenues has been an important, though not the only, factor in the rapid relocation of nearly all recent stars to other countries (in particular the USA). Culture has then had an influence on certain actors' approaches to the issue of copyright but it would be very wrong to say that such cultural or religious attitudes are relevant to today's Jamaican musicians (who are heavily influenced by high-consumption U.S. rap and hip-hop role models) or that such attitudes explain the lack of a functioning IPR regime.

In conclusion, Kingston has functioned as a world-class innovation center or creative milieu driven by a fragmented industrial structure that is highly competitive and embedded in a dense interactive and transactional social mode of production. Nonetheless, for a variety of reasons most of the value chain is outside the city and country, and the industry faces serious problems in retaining profits and securing investment necessary to improve its export potential.

Conclusion

It appears from these two different examples of successful music production centers that geographic proximity based on agglomeration in urban areas fosters intensified transactions and interactions that are crucial to commercially successful creative and innovative product development. It is important, however, to stress that rates of return to agglomerated economic activity and innovation vary considerably due to a variety of factors. Most crucial in the case of the music industry are the nature of firm level linkages between global and local actors, and the effectiveness of the intellectual property rights environment creativity is generated within. It seems that the particular organizational mix and structure found in a center of creativity may link directly with the IPR regime it is embedded within. Kingston can be seen in some ways as a model for what happens to a music industry's structure and *modus operandi* if an effective IPR regime doesn't exist. Stockholm, in contrast, is an example of the profitable effects of being fully integrated into the mainstream global IPR regime and associated industry configurations. The two cases demonstrate that, firstly, different positions within the present global political economy affect musical economies' abilities to protect IPRs and thus the ability to allow investments in creativity to be returned. Secondly, that if in the future the ability to commercially exploit copyrights gets significantly eroded (with, for instance, technological change) the music industry's institutions may well reconfigure or fragment along the lines of places such as Kingston.

The main focus of the study has been the ways in which internationally competitive product development links to the fullest possible returns on innovation for the production center and its industrial system and actors. In this area it has been shown that the two centers profit from their production at considerably different rates. On the face of it the simple lesson to be learnt from these two cases seems to be that higher degrees of firm-level integration into the wider global music industry confers major competitive advantages to a city's, or small nation's, music industry and the proportion of profits that return or remain in the center. Strong and institutionalized vertical intra- and inter-firm linkages to the global market have in the case of Stockholm had the effect of bringing Swedish products to global markets in a manner in which intellectual property is, as far as possible, guaranteed. Furthermore, exposure to the global has been crucial to the development of an extensive network of production and post-production services and firms which have had benefits for both domestic artists and export earnings. In contrast Kingston has remained largely outside the global music industry and therein an effective international copyright regime leaving it vulnerable in the past to exploitation by both the "majors" and by unscrupulous home grown criminals. We should, however, be careful not to neglect the fact that in both cases, and in particular in the case of Kingston, some observers have suggested a degree of separation and autonomy from global circuits may have been important in supporting a diverse and creative domestic music scene and related corporate structure. Many Western music fans romanticize the creative scenes associated with chaotic and unregulated industries such as Kingston's. Neglecting the fact that Jamaican musical products have long been internationalized and commodified, such people have suggested to us that greater internationalization and commodification of local culture for international consumption will suck the vitality and lifeblood out of the local scene; in short, don't tamper with something which has produced such great music. However, evidence from highly commodified and internationalized music scenes – e.g. New York, Los Angeles, London, Stockholm, Nashville, etc. – demonstrate that this need not be so. Commercial success can sit very well with, and indeed encourage, musical creativity and also provide financial rewards that places like Jamaica sorely need. It is in reality hard to romanticize or excuse the appalling and oppressive poverty, crime, and violence of the "scene" in Kingston.

In conclusion, it seems from these cases that the stronger the firm-level and institutional links between localized industry actors and multi-national corporations and the integration of the country into international IPR regimes the higher the rate of return (both financially and in terms of technical and innovation resources) to the local production center. If one were to think these cases worthy of generalization then it would seem that in the case of the music industry the spreading influence of the global majors should not necessarily be seen as the equivalent of the spread of cultural imperialism and centralized commercial exploitation as is so often the case. The majors are powerful actors in laying the ground for and financing not only musical products but the IPR issues surrounding them. In the case of these two production centers a complex set of relations between local and global scales exist and processes of linking the scales offer very real positive, as well as negative, possibilities for creativity, competitiveness, and profitability. Furthermore, whilst this chapter has concentrated on the music industry it is worth noting that the findings presented here could have implications for other cultural products industries, many of which also rely on effective IPR mediated through internationalized and corporatized value chains to get a fair share of the profits back "home." It is our viewpoint, then, that as with the music industry it is not only the quality of the creative milieus that leads to commercial success in cultural products industries but also the links between the local production system and international circuits of capital, distribution, and effective property rights. The recent rapid growth in the sizes and profitability of markets for cultural products suggests that these sectors may be of great help in economic development, and for smaller open economies with a proven track record of cultural product innovation the competitive position and rate of return in these industries must be carefully examined and supported by both the private and public sector.

Acknowledgments

The authors would like to thank Tommy Berg for his help with some of the material this chapter is based upon. This chapter is a revised version of a paper that appeared in *Environment and Planning A*, 34(10): 1833–54.

References

Alexander, P. (1996) "Entry barriers, release behavior, and multi-product firms in the popular music recording industry," *Review of Industrial Organization*, 9: 85–98.

Amin, A. and Thrift, N. (1992) "Neo-Marshallian nodes in global networks," *International Journal of Urban and Regional Research*, 16: 571–87.

Berg, T. (2001) *Funky Kingston: The Jamaican Music Industry in a Global Context*, Uppsala: Department of Economic History, Uppsala University.

Bourne, C. and Allgrove, S. (1995) *Prospects for Exports of Entertainment Services from the Caribbean: The Case of Music. A World Bank Report*, Washington, D.C.: The World Bank.

Burnett, R. (1990) *Concentration and Diversification in the International Phonogram*

Industry, Göteborg: Göteborg Studies in Journalism and Mass Communication No. 1, University of Göteborg.

—— (1992) "The implications of ownership changes on concentration and diversity in the phonogram industry," *Communication Research*, 19: 749–69.

—— (1993) "The popular music industry in transition," *Popular Music and Society*, 17: 87–114.

—— (1996) *The Global Jukebox: The International Music Industry*, London: Routledge.

—— (1997) *Den svenska musikindustrins export 1994–95*, Stockholm: ExMS.

Chang, K. and Chen, W. (1998) *Reggae routes: the Story of Jamaican Music,* Philadelphia, PA: Temple University Press.

Choi, C. J. and Hilton, B. (1995) "Globalization, originality, and convergence in the entertainment industry," in L. Foster (ed.) *Advances in Applied Business Strategy*, Greenwich, CT: JAI Press.

Christianen, M. (1995) "Cycles in symbolic production? A new model to explain concentration, diversity and innovation in the music industry," *Popular Music*, 14: 55–93.

Davis, S. and Simon, P. (1992) *Reggae Bloodlines – In Search of the Music and Culture of Jamaica*, New York: DaCapo.

Denisoff, R. and Bridges, J. (1982) "Popular music: who are the recording artists?," *Journal of Communication*, 32: 132–42.

Dolfsma, W. (2000) "How will the music industry weather the globalization storm?," *First Monday*, 5.; available online: <http://firstmonday.org/issues/issue5_5/dolfsma/index.html> (accessed 1 December 2003).

Eurostat (2001) *Information Society Statistics. Theme 4 – 4/2001*, Luxembourg: Eurostat/European Communities.

Forss, K. (1999) *Att ta sig ton: om svensk musikexport 1974–1999. Ds 1999: 28*, Stockholm: Fritzes.

—— (2001) *The Export of the Swedish Music Industry: An Update for the Year 2000*, Stockholm: Export Music Sweden.

Foster, C. (1999) *Roots Rock and Reggae: the oral history of reggae music from Ska to Dancehall*, New York: Billboard.

Galperin, H. (1999) "Cultural industries policy in regional trade agreements: the cases of NAFTA, the European Union and MERCOSUR," *Media, Culture and Society*, 21: 627–48.

Gilroy, P. (1987) *There Ain't No Black in the Union Jack*, Chicago: University of Chicago Press.

Hallencreutz, D. (2002) *Populärmusik, kluster och industriell konkurrenskraft: en ekonomisk-geografisk studie av svensk musikindustri,* Uppsala: Kulturgeografiska institutionen.

Harris, T. (1984) *Reggae Vibrations*, Kingston: T.B.O.J. Productions.

Hesmondhalgh, D. (2002) *The Cultural Industries*, London, Sage.

Hussey, D. and Whitney, M. (1999) *Bob Marley: Reggae King of the World*, London: Pomegranate.

IFPI (International Federation of the Phonographic Industries) (2001a) *The Recording Industry in Numbers 2001*, London: IFPI.

—— (2001b) *IFPI Music Piracy Report June 2001*, London: IFPI.

Jahn, B. and Weber, T. (1998) *Reggae Island: Jamaican Music in the Digital Age*, New York: Da Capo.

Johansson, S. (2000) *Time – företag I Stockholms Län: tillväxt och lokalisering av företagens verksamhetstyper 1990–1999*, Stockholm: TIME.

Kasaras, K. (2002) "Music in the age of free distribution: MP3 and society," *First*

Monday 7; available online: <http://firstmonday.org/issues/issue7_1/kasaras/index.html> (accessed 1 December 2003).

Kozul-Wright, Z. and Stanbury, L. (1998) *Becoming a Globally Competitive Players: The Case of the Music Industry in Jamaica. UNCTAD/OSG Discussion Paper No. 138*, Geneva: UNCTAD (United Nations Conference on Trade and Development).

Leyshon, A. (2001) "Time–space (and digital) compression: software formats, musical networks, and the reorganisation of the music industry," *Environment and Planning A*, 33: 49–77.

Lopes, P. (1992) "Innovation and diversity in the popular music industry, 1969–1990," *American Sociological Review*, 57: 56–71.

Malm, K. and Wallis, R. (1992) *Media Policy and Music Activity*, London: Routledge.

OECD (Organization for Economic Co-operation and Development) (2000) *Measuring the ICT Sector*, Paris: OECD.

OPM (Office of the Prime Minister) (1996) *Jamaican National Industry Policy*, Kingston: Government of Jamaica.

Peterson, R. and Berger, D. (1975) "Cycles in symbolic production: the case of popular music," *American Sociological Review*, 40: 158–73.

—— (1996) "Measuring industry concentration, diversity and innovation in popular music," *American Sociological Review*, 61: 175–8.

Pfahl, M. (2001) "Giving away music to make money: independent musicians on the internet," *First Monday* 6; available online: <http://firstmonday.org/issues/issue6_8/pfahl/index.html> (accessed 1 December 2003).

PIOJ (Planning Institute of Jamaica) (1997) *Economic Update and Outlook*, Vol. 1, No. 4, Kingston: PIOJ.

Porter, M. (1990) *The Competitive Advantage of Nations*, New York: The Free Press.

Potash, C. (1997) *Reggae, Rasta, revolution: Jamaican Music from Ska to Dub*, London: Prentice Hall International.

Power, D. (2002) "The 'cultural industries' in Sweden: an assessment of the place of the cultural industries in the Swedish economy," *Economic Geography*, 78: 103–127.

—— (2003) *Final Report: Behind the Music – Profiting from Sound: A Systems Approach to the Dynamics of Nordic Music Industry*, Oslo: Nordic Industrial Fund – Center for Innovation and Commercial Development.

Saakana, A. (1980) *Jah Music: The Evolution of the Popular Jamaican Song*, London: Heinemann Educational.

Sadler, D. (1997) "The global music business as an information industry: reinterpreting economies of culture," *Environment and Planning A*, 29: 1919–36.

Scott, A. (1999a) "The U.S. recorded music industry: on the relations between organization, location, and creativity in the cultural economy," *Environment and Planning A*, 31: 1965–84.

—— (1999b) "The cultural economy: geography and the creative field," *Media, Culture and Society*, 21: 807–17.

—— (2000) *The Cultural Economy of Cities: Essays on the Geography of Image-Producing Industries*, London: Sage.

Shapiro, D., Abercrombie, N., Lash S. and Lury C. (1992) "Flexible specialisation in the cultural industries," in H. Ernste and V. Meier (eds) *Regional Development and Contemporary Industrial Response*, London: Belhaven.

Stephens, M. (1998) "Babylon's 'natural mystic': the North American music industry, the legend of Bob Marley, and the incorporation of transnationalism," *Cultural Studies*, 12: 139–67.

Storper, M. and Salais, R. (1997) *Worlds of Production: The Action Frameworks of the Economy*, Cambridge, MA: Harvard Universtiy Press.

Taylor, M. and Towse, R. (1998) "The value of performers" rights: an economic approach," *Media, Culture and Society*, 20: 631–52.

Wallis, R., Baden-Fuller, C., Kretschmer, M. and Klimis, G. (1999) "Contested collective administration of intellectual property rights in music: the challenge to the principles of Reciprocity and Solidarity," *European Journal of Communication*, 14: 5–35.

Wallis, R. and Malm, K. (1984) *Big Sounds from Small Peoples: The Music Industry in Small Countries*, London: Constable.

Watson, P. (1995) *The Situational, Analysis of the Entertainment (Recorded Music) Industry*, Kingston: The Planning Institute of Jamaica.

14 Cultural industry production in remote places

Indigenous popular music in Australia

Chris Gibson and John Connell

This chapter discusses cultural production in Australia, focusing on a case study of Indigenous popular music in remote parts of Australia. It is partly intended as a counterpoint to the thrust of much research on the geography of cultural industries, which focuses on agglomerations or clusters of activity in districts of major western cities. It is concerned with cultural production in some of the most remote parts of the world, and in circumstances of extreme socio-economic disadvantage. The chapter therefore seeks to examine the structure of cultural production in scattered, distant places that are vastly different from the conventional urban clusters, and explore how recent technological and political changes provide opportunities for more dispersed or decentralized activities. Cultural industries are usually most successful when production agglomerates in urban areas, particularly major metropolitan centers (Connell and Gibson, 2003), yet the creative activities (music making, writing, painting, etc.) upon which cultural industries rely take place across much wider distances and often dispersed contexts that are far from being hubs of capital and investment. The extent to which cultural activities in such locations may be transformed into export-earning industries is the focus of this chapter. It draws together earlier research projects on Indigenous production of popular music (Gibson, 1998; Connell, 1999; Dunbar-Hall and Gibson, 2004). These projects involved interviews with producers, managers, promoters and musicians, and analysis of production, employment and business location data. Insights drawn from this case study shed light on both the policy implications of cultural production by Indigenous groups in other countries (for example, in Canada and the United States), and the theoretical implications of creative workers being physically and economically distant from recognized centers of cultural production.

Australia is itself physically distant from most other parts of the world, with the exception of the Asia-Pacific region. Recent changes in the economic structure of global cultural production and increasing technological capacity has of course changed the ways in which all locations – close and afar – are incorporated into the distribution systems of entertainment corporations. Thus Australians are now more likely than ever to consume similar music, films, books, clothes and sports as people in other parts of the world. Youth subcultures, changing demographic structures, the commodification of cultural diversity (most evident in food consumption, but also in music) and greater mobility (partly through the influence of tourism) have resulted in increasing complexity of cultural production. Global cultural styles, fashions and subcultures, many originating in places like London or New York, are almost immediately present in Australian capital cities, and have been appropriated and altered in local circumstances to create hybrid creative forms.

Tracing the flow of creativity in the opposite direction, Australian cultural production is critically positioned in an international sense. While transformed by global influences and technologies, many Australian creative pursuits are still in essence local ventures, in that they seek, at least initially, mainly domestic audiences. However, due to Australia's relatively small population, certain Australian cultural industries have become increasingly export-orientated since the 1980s. There are numerous examples of where this has been successful. Domestic firms have generated product for Anglophone markets in music, film, TV and publishing, many succeeding due to their mix of international marketability (appealing to global tastes and norms), and distinct local inflections. Australian films such as *Mad Max*, *Crocodile Dundee*, *Priscilla: Queen of the Desert*, *Babe* and *Moulin Rouge!* were highly successful. More recently, Sydney has become a location for offshore U.S. film production, including the *Star Wars* and *Matrix* trilogies. Australia has also been a major supplier of international music. A well-established career path for Australian musicians encourages relocation overseas to "crack" North American and European markets after establishing local reputations (sometimes as actors in TV soap operas that have had export success too). Successful examples have included Rolf Harris, the Bee Gees, AC/DC, Air Supply, INXS, Crowded House, Men at Work, Kylie Minogue, Nick Cave and the Bad Seeds, and more recently Savage Garden, silverchair, Holly Valance and Natalie Imbruglia.

Almost all of this export-orientated cultural production has emerged from the largest cities in Australia, despite films often being made inland. Like the population of Australia, commercial cultural production has hugged the coastline. Moreover, despite films and television series such as *Flying Doctors* being shot in the inland, these cultural industries have involved Europeans at every stage of the process, with very little Indigenous Australian content. However, there is a very different structure of cultural production in parts of remote and inland Australia, only tangentially linked to the wider national structure of cultural industries, and one that has involved both art and music. It is in these contexts that more substantial insights may be gained into the influence of geography on cultural production.

Remote cultural production in Australia: a colonial context

A limited number of studies have shown how cultural industries are bound up in transformations of rural areas and small urban places (e.g. Kneafsey, 2001; Gibson, 2002; Gibson and Connell, 2003). This has contributed to a breaking down of distinctions between "cities" and "rural" spaces, which have hitherto tended to be perceived as separate space-economies (see MacLeod, 2001). Much work on cultural industries has tended to examine activities within localities, and underestimate inter-regional flows of culture and capital (Gregson *et al.*, 2001). Moreover, commodity chains for economic activities are also *geographies* of production and consumption; they necessarily function in and across space, and are constituted through spatial relations (Leslie and Reimer, 1999). The contexts of cultural industry activities are likely to vary enormously for commodities as they are produced, distributed and consumed in different places.

This chapter extends this discussion to a specific set of locations of cultural production and consumption where colonial relations continue to underpin all activities, including those in the cultural industries. In this case creative production has emerged in a "fourth world" context – remote Indigenous Australia – that is quite removed from the inner-city, upwardly mobile social setting of most cultural industries.[1] The present study focuses on the Northern Territory, where some 50,000 people are scattered over 1.3 million square kilometers, and many of the Indigenous population are extremely remote from the two main urban centers. This setting is thoroughly impacted by the continuing effects of colonialism; across health, economic and social indicators, Indigenous Australians remain severely disadvantaged in comparison with the non-Indigenous population. In 2000 the official Aboriginal unemployment rate – which itself underestimates unemployment in remote areas – was 18 percent, compared to 7 percent for the non-Aboriginal population; life expectancy for Indigenous males is currently 56 years, and for females 63 years, compared with national figures of 77 and 82 respectively (Howitt, 2003). Such extensive disadvantage, alongside the dispossession of traditional land and the impacts of paternalist assimilation policies during the twentieth century, has produced a very different context for cultural industries. The Indigenous population is marginal geographically, economically and socially yet their cultural production had become very significant. Remote from major metropolitan centers, where cultural industries have tended to cluster, questions of employment, empowerment and the persistence (or consequent resistance and destabilizing) of colonial power relations become critical. In Aboriginal contexts notions of "economic development," as well as "empowerment" and "self-determination," are highly multivalent terms (Gibson, 1999) reinterpreted and redefined depending on the contexts within which they are articulated.

The emergence and importance of cultural industries have particular resonance for remote Indigenous communities in Australia. Cultural industry participation has been promoted as an avenue for employment, Indigenous self-determination and independence from welfare reliance, but particularly for remote Indigenous communities with dispersed, small populations and little access to formal economic flows. Cultural industries have also been popular with policy-makers because the texts that are produced – music releases, films, television programs, books, art – are themselves capable of expressing Indigenous perspectives, and are thus crucial to the agenda of Aboriginal self-determination that gained pace in the 1970s and 1980s in Australia, as official federal policies of self-determination encouraged Aboriginal groups to maintain culture, including traditional languages and tribal law. This mix of economic and expressive elements made cultural industries obvious targets of Aboriginal economic and social development programs.

In the mid-1990s, the peak national Indigenous bureaucracy, the Aboriginal and Torres Strait Islander Commission, developed complementary National Tourism and Cultural Industry Strategies for Indigenous communities (ATSIC, 1997a, 1997b). These attempted to identify opportunities for Indigenous Australians in tourism and the cultural industries, including linkages between cultural producers and tourism providers, and models of Aboriginal enterprise development. These strategies were one response to the recommendations of the Royal Commission into Aboriginal Deaths in Custody conducted in the late 1980s and early 1990s, which sought to uncover a range of social, economic, legal and political factors contributing to high levels of incarceration of Aboriginal people in Australia's prisons, and subsequent deaths while in correctional facilities. Cultural industries were identified as priority areas for economic development, particularly because Indigenous musical, artistic and literary expressions might also improve the visibility and acceptance of Indigenous peoples within wider Australian society, and strengthen Aboriginal cultural values.

The most significant manifestation of the policy shift towards self-determination and the promotion of Indigenous cultural expressions has been the rise of a network of both urban and remote Indigenous broadcasting agencies, funded in part through the Federal Government's Broadcasting for Remote Aboriginal Communities Scheme (BRACS). This network involves over 100 broadcasting agencies based in remote communities and other centers with significant Indigenous populations. These agencies work across cultural sectors, but have become best known for television and film production, which has had particular significance for "cultural maintenance" (Michaels, 1986; Hinkson, 2002), and music, with several establishing professional recording studios as well as television-production facilities. The most important of these has been the Central Australian Aboriginal Media Association (CAAMA), established in Alice Springs in the Northern Territory in 1980, over 1,000 kilometers from the nearest metropolitan center. CAAMA began radio broadcasting in 1980, its "footprint" covering the vast center of Australia. Video recording began in 1983, and television broadcasting (through Imparja Television) in 1988. CAAMA has specific objectives that relate directly to its role as a recorder and distributor of Aboriginal music, and has sought to use popular music as a means of spreading information, celebrating Aboriginal cultures, and addressing particular problems in Aboriginal communities.

CAAMA gave both new legitimacy and high-quality production facilities to Aboriginal musicians and groups. It has released a vast amount of cassettes and compact discs by remote community bands, mostly from the Northern Territory. At the same time, the growth of Aboriginal radio stations and programs in remote areas, and in major cities and towns, meant that performers had new means to communicate with their own people, and access to new, wider

Aboriginal and non-Aboriginal audiences. These services, established with an explicit mandate to broadcast Indigenous music, have been crucial to the expansion and viability of Indigenous music in Australia. This is particularly so given the persistent reluctance of most commercially orientated radio stations to broadcast the work of Indigenous musicians not aligned with major record companies.

"Deadly" sounds:[2] the emergence of indigenous popular music scenes

The flow of public funding into Indigenous creative industries in the 1980s and 1990s mirrored an expansion in the commercialization and wider distribution of Aboriginal creative expressions. Aboriginal art in particular became a major export earner in the 1980s and 1990s, with exhibitions in Europe and North America, and the rapid commercialization of Aboriginal designs, particularly for tourist souvenirs, but also for overseas markets (Myers, 2003). Indigenous musicians and dance companies also began to score success both within Australia and overseas, including musicians such as Yothu Yindi, Archie Roach and Tiddas, rock operas such as Bran Nu Dae and Sydney's Bangarra Dance Theatre Company. Such successes led to perceptions among policy makers and promoters of economic development that Indigenous communities had a particular comparative advantage in "culture" – a source of unique images, sounds and movements capable of commodification.

Tracing Indigenous involvement in popular music entails engagement with the wider contexts of commerce, technology and social change that transformed virtually all production and consumption of music during the nineteenth and twentieth centuries. Music became a more influential international cultural influence with the availability of mass-produced, printed sheet music, and subsequently with the advent of the phonogram, pre-recorded music and radio. Musical styles such as minstrelsy, vaudeville and Anglo-Celtic balladry grew out of their North American and European roots and developed into polished products alongside other emerging sounds such as Dixieland, Tin Pan Alley, swing and big-band music. These forms of music were distributed, replicated and mimicked in a variety of distant geographic contexts, initially through touring groups, and later through the influence of radio. Many were absorbed by Aboriginal musicians in rural and remote communities throughout Australia during the early twentieth century; traces remain in the strong Aboriginal traditions of bush ballads, gospel-influenced hymns and "gum-leaf orchestras" (Dunbar-Hall and Gibson, 2004). Despite the popularity of these early forms of Indigenous popular music within their own network of communities, the attitudes of then assimilationist Australia severely limited the boundaries within which Aboriginal musical expression were received by wider audiences. Few artists were ever recognized in urban areas (with rare exceptions, including country musician Jimmy Little). Aboriginal music scenes were almost wholly divorced from those elsewhere in Australia.

The most dramatic shifts in the nature of Aboriginal popular music have occurred during the past 25 years, reflecting official moves away from paternalism and assimilation towards self-determination, but also the arrival of the electric guitar, drum kits and cassette players. A plethora of Aboriginal musical artists have emerged, from a variety of geographical contexts, writing songs and recording albums that reflect the diversity of experiences in different regions of Australia. Contemporary Indigenous popular music production is grounded in Indigenous music "scenes" across urban, rural and remote areas. In metropolitan centers Indigenous musicians have absorbed the influences of global musics, particularly hip hop, R&B and dance styles. More than other Australian cultural industries, the vast majority of Aboriginal music production has emerged from remote and scattered circumstances throughout the north – in central Australia, the Kimberley region, the Top End of the Northern Territory and Cape York. This is particularly the case for reggae, country, rock and gospel music.

Many of the names of bands and recordings chosen by Aboriginal musicians reflect the geography of Indigenous popular music in Australia, and articulate connections to traditional country (tribal lands). Bands commonly adopt their home community name, a feature of their nearby environment or a unique aspect of their local culture for their own name (Dunbar-Hall and Gibson, 2004), hence the Titjikala Desert Oaks Band, The Ltyentye Apurte Band (who play, in their own words, "desert surf music") or the North Tanami Band (well known for their album *Travelling Warlpiris*, a case of sardonic appropriation by Warlpiri-language speakers of an Anglo-American music supergroup of similar name). This process signifies not only the geographic roots or origins of the musicians involved, but affirms a sense of affiliation with traditional country. Aboriginal music thus both derives from, and gives rise to, particular geographies, as the popularity of different styles of music in different centers and regions of remote Australia supports various local bands who mix outside influences with local concerns in new syncretic expressions.

The release of Indigenous recordings increased during the 1980s as a result of a number of factors: establishment and funding of remote broadcasting agencies; the rise of a pan-Indigenous audience for Australian Aboriginal music; increasing social networks between remote communities opened up by the greater availability of motorized transport; and the new regional activities of land councils, tribal organizations and community associations. The increasing presence and distribution of Indigenous music through these networks has also clearly benefited from changing technology, in particular the advent of cheaply recorded and distributed audio cassettes that could be played on battery-operated portable equipment (in environments where vinyl records could simply melt). In communities where the availability and consistency of electricity supply is dubious, where populations are transient, and where resources for high-quality recording sessions are non-existent, the growth of a vibrant "cassette culture" (Manuel, 1993) allowed the diffusion of musical products to continue, though such means of communication as the print media remain less pervasive. The production and sales of Indigenous music continue to reflect the dominance of cassettes as a format.

Interest in Indigenous music, and the establishment of a wave of new Aboriginal bands, was spurred on by the domestic and international success of Yothu Yindi in the early 1990s, followed to a lesser (commercial) extent by Christine Anu, Warumpi Band, Archie Roach, Troy Cassar-Daley and Tiddas, though only Yothu Yindi and Warumpi Band came from very remote communities. Interest from major record companies peaked in 1994 with band, solo artist and compilation releases from Phonogram (now part of the Universal Music Group), RCA (now part of the Bertelsmann Music Group), EMI and Festival/Mushroom (Australia's "major" label, owned by Rupert Murdoch's News Corporation). At that time the majors were imbricated in an extensive "love affair" with independent record companies in Australia. In Australia the majors were exploring the possibilities of niche marketing efforts in the wake of the emergence of dance/techno, "indie" music and other youth subcultures, and in reaction to declining sales figures for "mainstream" rock artists. For the major companies Indigenous music was another new "niche" capable of being transformed into products with widespread marketability.

The interest of major record companies ushered in a flurry of activity to sign "the next Yothu Yindi" (after the 1991 success of their hit "Treaty"), and optimism grew amongst Indigenous musicians for greater commercial rewards through corporate support. Above all, this period created new and very different linkages between metropolitan cultural gatekeepers (record company executives, A and R staff) and those in the most distant and vulnerable communities in Australia. Indigenous musical voices, alongside those emerging in art and literature, entered international mediascapes on a scale hitherto unseen. While political or postcolonial agendas were peripheral to the intentions of major record companies when distributing Indigenous music, commodification nonetheless created economic networks and lines of cultural influence that transcended distance.

A remote indigenous musical economy

Seven years on from the boom in Indigenous musical production, none have quite emulated the commercial success of Yothu Yindi, though many have found popular appeal amongst certain audiences, such as Troy Cassar-Daley in the country scene. Beyond such individual examples, the amount of Indigenous music being released has fallen, and is now largely made up of releases by Indigenous media organizations in remote areas of the country. As the majors lost interest most contemporary Indigenous music again became mostly local in its production and consumption, and thus mostly remote in origin, in contrast to occasional engagements with circuits of corporate entertainment capital. Rates of commercial release from remote communities have in themselves grown (in part a reflection of cheaper and better-quality recording gear), but have only exceptionally penetrated urban markets.

The remoteness of Indigenous communities and their music has given a particular character to the economic dimensions of the scenes where most contemporary indigenous music is now made, and the cultures of consumption that support them. Wider issues of the participation of Indigenous people in remote, regional economies (see Fisk, 1985; Crough, 1993) are also present in the cultural industries, including links between social welfare and small-scale private-sector activities, varied lifestyles that involve elements of waged production and subsistence, and the impacts of remoteness on regional economic formations.

The conditions of production, performance and distribution for Aboriginal musicians are cut across by geography. Issues like racism, gender relations of production and employment, lack of company support and inadequate promotion (Dunbar-Hall and Gibson, 2004) are exacerbated by distance. Access to industry contacts and opportunities can be as much a function of proximity – of being "in the right place at the right time" – as of talent. That many Aboriginal musicians come from, and sing about, what non-Aboriginal people would perceive to be "remote" places limits opportunities for those seeking success in the mainstream, because the music industry (and especially venues) remains highly concentrated in Sydney and Melbourne.

Musicians, managers and promoters of Indigenous music have all emphasized how large distances discourage touring even between Indigenous communities; it is difficult to get a coordinated show on the road in remote Australia, particularly if that tour involves lengthy travel on unsealed roads and uncertain weather conditions. Difficulties in promoting Indigenous releases and tours across great distances – even to the point of, in the case of one band having to barge public-address systems and other audio equipment between coastal communities – prevent further expansion of markets for Indigenous music beyond radio, which remains the most important means of dissemination. In contrast to the now almost monopolistic control of music distribution in Australia (with one firm, Entertainment Distribution Company (EDC), undertaking centralized manufacturing and distribution for Sony, EMI and Warners), networks of distribution for Indigenous recordings are themselves fluid and not always permanent. CAAMA undertakes formal distribution from its Alice Springs studios to record stores around Australia, and to communities who stock a regular collection of cassettes in their community general stores. A few other small-scale labels operate in a similar manner; Skinny Fish Music, based in Darwin, is one of the more successful throughout the Northern Territory, with both the Saltwater Band, from Galiwin'ku, and Narbarlek Band, from Manmoyi. Agreements to distribute product are dealt with by a small group of individuals who negotiate directly with storekeepers of community stores, while band tours are staged as one-off ventures, taking in a series of Indigenous communities (such as Lajamanu, Barunga, and Maningrida in the Northern Territory). Other bands simply undertake the funding of production, distribution and organization and management of tours themselves. Such involvement, while ensuring Indigenous control over the terms of engagement with the formal economy, is extremely labor intensive, and only likely to be undertaken intermittently, particularly as many band members maintain links to communities and traditional country that require substantial periods of time at home. Certain elements of ceremonial life in remote communities clash with expectations of the Australian music industry. As Neil Murray, from the Warumpi Band, has stated:

> It's hard for any band coming from a remote area. Coming from the bush, it was a bit hard for us to give a serious commitment to being down south in the cities because we'd only be down there for so long and everyone would become really homesick. It's hard to reconcile [traditional] values with the commitment required to be a successful band. If you're booking a tour, you book it three months ahead, and then suddenly something happens: somebody dies and then someone can't go because they've got sorry business [Aboriginal funeral ceremonies]. You're blowing out gigs and everyone down south loses faith in you. They think you're unreliable.
>
> (Quoted in Mitchell, 1996: 28)

The costs of mounting tours of capital cities and country towns – including travel, accommodation, haulage and PA hire – prevent many artists from remote areas being able to effectively promote their work elsewhere. This barrier is made more difficult for artists who have social responsibilities that also prevent them leaving for any length of time, and in a part of the world where there can be no economies of scale.

Aboriginal musicians have responded to these challenges in a number of ways. Many Aboriginal groups have resolved to only play at Aboriginal festivals, where the cultural agendas of survival and revival are prioritized. Due to the need to access large Indigenous audiences, and partly because of the hostility directed towards many Indigenous musicians from other venues in predominantly non-Aboriginal remote mining towns, a distinct network of venues and spaces for live music performances has emerged throughout Aboriginal communities. The links in touring networks mirror the socio-linguistic connections between communities over a wide remote region, as in the matrix of Top End communities of the Northern Territory who speak Yolngu language (Corn, 2002), and tours through this by community bands such as the Sunrize Band and Blekbala Mujik.

Some bands survive financially in this way and build up Indigenous audiences without "crossing over" into capital city markets:

> There is a network for Aboriginal bands if they can get themselves organized. They can play their communities. That's how the Warumpi Band used to do it. We couldn't get a gig in Alice Springs for years in the pubs. They reckoned we were too much trouble. So we had to play in people's backyards, play in a hall or out at somebody's basketball court in the community, but we got by.
>
> (Neil Murray, quoted in Mitchell, 1996: 28)

The economics of music performance now also favor festivals. Rather than following extensive metropolitan and rural touring networks, having to negotiate racism from non-Aboriginal venue managers and trying to earn an income through numerous smaller (and poorly paid) pub gigs, bands can resist "saturating" their audiences with performances, and pick up higher returns in the long run by performing occasionally at festivals (also reducing on-going travel and accommodation costs). For many in this situation, music has simply ceased being a full-time occupation, but longer careers continue on an intermittent basis, as with Warumpi Band and Coloured Stone, who have had musical careers lasting over 20 years.

Self-production and promotion have become options for some. For example Jason Lee Scott sells his CDs via the internet, after failing to secure a distribution deal; other musicians and Aboriginal companies are doing likewise, including CAAMA which now has an extensive website and catalogue ordering system. As the Aboriginal musician Kev Carmody argued:

> changing technology is making it easier – it's now more portable, cheaper, people are writing and producing their own stuff, whereas once the record company got you signed up, gave you $50,000 to record an album and you're trapped for life. Now, the whole situation has changed and people are able to do their own thing to some extent.
>
> (1996 interview with author)

Access to technology, electricity and appropriate skills in remote communities still restricts the empowering possibilities of recording and distribution. Nonetheless, the potential exists for further decentralization of the means of production throughout Aboriginal communities, as digital technology becomes cheaper and easier to use, particularly if public funding for computer resources in communities is improved.

Limited Indigenous participation in other parts of the music industry makes it very difficult to successfully manage all aspects of production beyond the established infrastructure in the music industry. For Kev Carmody, "there simply aren't enough trained Aboriginal people. For example, about 5 or 6 years ago I tried to make a promotional music video which was completely black – employing black people all the way – I couldn't find a DOP [Director of Photography]" (1996 interview with author). Many Aboriginal musicians have long stressed the need for an overarching umbrella organization – an Aboriginal musicians' union of sorts, as well as a networking agency for Aboriginal performers, technicians and support staff in the music industry. Songlines Aboriginal Corporation in Melbourne and CAAMA in Alice Springs both perform these roles to some extent, with training programs and stated agendas to improve the situation of Aboriginal musicians, yet their activities are ultimately constrained by funding and geography – tied as they are to two very different places and musical scenes.

An international market?

Following the success of tours to Europe, Asia and North America by Aboriginal groups such as Yothu Yindi, Blekbala Mujik and the Bangarra Dance Theatre, attempts to promote Aboriginal music overseas have increased. Such attempts are partly fueled by an increasing perception of the existence of more sympathetic audiences overseas, and ultimately by greater demand for Aboriginal music. For Aboriginal country musician Bobby McLeod, "You're not sort of thought about in this country in a positive way, but people overseas look at us like we're magicians or masters of something. It makes you want to do more for your culture. It's a pity [Aboriginal culture and people] are not respected as much in this country as in other countries' (quoted in Condie, 2001: 33). Driven by similar concerns, Blekbala Mujik signed North American distribution deals, and at various times toured Europe (including a performance at the 1997 WOMAD Festival in the Canary Islands); Yothu Yindi continue to sell out concerts throughout Europe; Tiddas toured America after their albums were distributed there by U.S.-based record companies; and, despite lack of interest in their native Melbourne, Blackfire have twice toured China:

> Our performance on China Central Television was an outstanding success. We've been basically ignored by the mainstream music industry in Australia and to reach an audience like that [est. 20 million] was just something other artists in Australia dream about. This was our second tour of China and the response from the Chinese people has been overwhelming. It's ironic that the Asian audiences have taken us into their hearts whereas in Australia it feels like we've been hitting our heads against the wall . . . if we can crack the market in Asia and become popular over there, I guess that just increases the focus for other Indigenous bands in Australia to try their luck. It's trying to create an awareness somewhere else to try and expose ourselves back in Australia.
>
> (Quoted in Evorall, 1999: 31)

Walbira Watts, director of Emu Tracks, an Aboriginal-owned event promotion company, confirmed the considerable latent demand for Aboriginal music overseas, but was skeptical of automatic success for musicians exploring this path, emphasizing economic and logistical concerns:

> I truly think there is demand [overseas]. The major problem though is the cost of airfares to send artists from Australia overseas and it isn't that easy to access funding to help this side. By the time an overseas festival has come up with airfares, accommodation and meals for the artists, there is often only a very minimal fee or often none at all to be offered, which brings up a whole set of problems. For example, many Aboriginal people are more than keen to go and share culture with overseas audiences regardless of the fees, but the artists themselves are often the main breadwinner for their family.
>
> (Quoted in Howes, 2001: 19)

International markets can be much more lucrative, simply due to larger populations. Even if Aboriginal musicians only captured a small percentage of the American and European markets, these would return much larger sales than even substantial success in Australia. However, while these larger markets are powerful lures for musicians, "international popularity exacts a price in terms of the emphasis on timeless and de-territorialized imagery which typically softens or submerges the political and social concerns of the writers' (Hollinsworth, 1996: 63). Amy Saunders experienced this in Tiddas' tour of the United States:

> America had a way of making us feel out of place. They were saying "now don't sing about Aboriginal issues because nobody here knows what's going on. And don't talk like this and don't talk like that." It's like well if we fucking don't do any of that, then who are we supposed to be? The Indigo Girls? I don't think so. They thought our only replies should only ever be "yes" and "okay" . . . they weren't prepared for three women who were going to say "no we're not wearing lycra and bikini tops and flouncing around with ridiculous makeup on."
>
> (Quoted in Hunter, 1996: 12, and Smith, 1996: 20)

Overseas touring and promotion merely emphasize the friction of distance already discussed, with higher costs, longer periods away from home and greater risks. Troy Cassar-Daley, for instance, traveled to Nashville for its annual Fan Fair, recorded an album and gained support from his record company, Sony, to launch a promotional campaign in the United States. However, as Warner-Chappell's Gina Mendello, who managed Cassar-Daley, argued:

> What you do here in Nashville is you get in a queue. You might get signed, but then they say, okay, we've got a few other things to release first. Priorities. It's a huge commitment of time, energy and money on behalf of a record company. And with that attitude, everybody's very careful, watching their budgets . . . You gotta have a manager, a publicist, you gotta get a bus, a staff, go on tour – and that's not something you can just pop into town and do. You gotta work on promotion, you gotta work on everything . . . and the thing is, you have to be here.
>
> (Quoted in Walker, 2000: 293)

Tensions thus persist between Indigenous desires to succeed in the music industry, the career paths they are generally expected to follow, domestic cultural constraints and the tyrannies of distance. In music, as in other cultural industries, creative producers move through different networks as their careers develop, and also perceive in quite different ways the networks that they may want to move through, in order to satisfy the desire to create expressions, as well as receive monetary reward. In the Australian music industry more generally, musicians have long sought to achieve success overseas, in part because the small market in Australia cannot support major incomes for even quite successful artists, but also because such journeys authenticate the reputation of the musicians. For Indigenous musicians, desires to transcend isolation are mediated by remoteness, which combined with the small, scattered nature of Aboriginal populations, prevents economies of scale upon which most cultural industry activities are based. Compounding this is institutionalized marginalization that has largely excluded Indigenous people from gatekeeping positions within the industry.

Popular music – a post-colonial cultural industry?

The shifting successes and particular geographies of Indigenous music production and consumption demand new ways of interpreting economic activities. The emergence of an Indigenous popular music industry in Australia demonstrates that remote cultural production is possible, while contemporary technologies make this more viable than in previous eras. Indigenous musical economies, though, are largely low-key and informal, intricately connected with community life and social networks. Similarly Indigenous music exists without the resources capable of developing extensive marketing campaigns, the capital to fund comprehensive tours (to "work new territories"), or the market power to ensure that metropolitan stores and radio stations feature Indigenous music on their shelves and in their high-rotation programming. Even acquiring regional success is costly and difficult.

Despite the economic, cultural and geographical constraints to Indigenous cultural industries suggest creative production has important empowering dimensions. The networks of small-scale activities that sustain Indigenous music performance and production across remote regions represent versions of what a more genuinely post-colonial remote regional economy might look like, even if perceived as vulnerable from a western economic development perspective. With substantial Indigenous control over production, Indigenous creativity is deeply embedded in the diverse practices that sustain cultural identities amongst Indigenous communities in the face of various forms of dispossession and marginalization. Much successful Indigenous involvement in art and music comes from remote contexts – where access to infrastructure and mainstream employment opportunities are highly constrained – but where traditional values remain strong.

Cultural occupations also offer flexible employment conditions in community contexts where work duties ideally complement, rather than disrupt, other daily responsibilities and obligations (such as the maintenance of traditional ceremonies, on-going land-management activities and participation in community decision-making arenas). A commitment to the principle of Indigenous self-determination, a basic premise for any genuinely post-colonial economic strategy, requires new perspectives and flexibility on the part of government, industry and policy-makers over what is meant by "economic development" and "employment," and the conditions under which work is carried out. If

Indigenous communities are encouraged to increasingly determine their own affairs, including the maintenance of tradition, language and customary practices, being able to decide where, when and whether to work within a formal economic sphere remains crucial. The cultural industries are one of a tiny number of possibilities for delivering a relatively flexible environment for engagement with formal spheres of work. Moreover, indigenous control over cultural production enables Aboriginal people to represent themselves through a mix of images, sounds and words in the media. The value of promoting Indigenous participation in cultural industries extends far beyond, and is independent from, purely "economic" benefits.

In remote Australia, for its Indigenous occupants, the cultural industries are genuinely cultural, in that they actively support indigenous culture (though in no sterile, reified form). At the same time they allow an engagement with the wider regional and global economy that no other economic activity has been able to provide on a similar scale and within Aboriginal communities. Moreover, art and music especially have created and allowed a new respect for Aboriginal culture, which in turn has stimulated movements towards greater self-reliance. Cultural industries in remote Australia may have few links to, or parallels with those in urban Australia, let alone in distant world cities, yet perhaps more than in almost any other context they are the expressions of communities that have long been grounded in particular places.

Acknowledgments

This research was funded through grants from the Australian Research Council (ARC) and the University Research Support Program, University of New South Wales. We would like to acknowledge the advice of Peter Dunbar-Hall and Vanessa Bosnjak in various stages of this research.

Notes

1 The 2001 National Census enumerated 410,000 Indigenous people in Australia, or 2.2 percent of the total population (Howitt, 2003). This figure is much higher (25 percent) in the Northern Territory.

2 The term "deadly" is Aboriginal English for "good" or "fantastic." It is widely used in association with music: "deadly tunes," "deadly beats," "deadly grooves," etc. Australia's annual national Indigenous music awards are thus called the "Deadly Awards."

References

ATSIC (1997a) *National Aboriginal and Torres Strait Islander Cultural Industry Strategy*, Canberra: ATSIC.

—— (1997b) *National Aboriginal and Torres Strait Islander Tourism Industry Strategy*, Canberra: ATSIC and the Office of National Tourism.

Condie, T. (2001) "McLeod's album is a paradox," *Koori Mail*, 19 September: 33.

Connell, J. (1999) "My Island Home: the politics and poetics of the Torres Strait," in

R. King and J. Connell (eds) *Small Worlds, Global Lives: Islands and Migration*, London: Pinter.

Connell, J. and Gibson, C. (2003) *Sound Tracks: Popular Music, Identity and Place*, London and New York: Routledge.

Corn, A. (2002) "*Burr-Gi Wargugu ngu-Ninya Rrawa:* Expressions of ancestry and country in songs by the Letterstick Band," *Musicology Australia*, 25: 76–101.

Crough, G. (1993) *Visible and Invisible: Aboriginal People in the Economy of Northern Australia*, Darwin: North Australia Research Unit and the Nugget Coombs Forum for Indigenous Studies.

Dunbar-Hall, P. and Gibson, C. (2004) *Deadly Sounds, Deadly Places: Contemporary Aboriginal Music in Australia*, Sydney: UNSW Press.

Evorall, T. (1999) "Blackfire sweeps Chinese market," *Koori Mail*, 19 May: 31.

Fisk, E. K. (1985) *The Aboriginal Economy in Town and Country*, Sydney: George Allen and Unwin.

Gibson, C. (1998) ""We Sing Our Home, We Dance Our Land": Indigenous self-determination and contemporary geopolitics in Australian popular music," *Environment and Planning D: Society and Space*, 16: 163–84.

—— (1999) "Cartographies of the colonial and capitalist state: a geopolitics of indigenous self-determination in Australia," *Antipode*, 31: 45–79.

—— (2002) "Rural transformation and cultural industries: popular music on the New South Wales Far North Coast," *Australian Geographical Studies*, 40: 336–56.

Gibson, C. and Connell, J. (2003) "Bongo Fury: tourism, music and cultural economy at Byron Bay, Australia," *Tijdschrift voor Economische en Sociale Geografie*, 94: 164–87.

Gregson, N., Simonsen, K. and Vaiou, D. (2001) "Whose economy for whose culture? Moving beyond oppositional talk in European debate about economy and culture," *Antipode*, 33: 616–46.

Hinkson, M. (2002) "New media projects at Yuendemu: inter-cultural engagement and self-determination in an era of accelerated globalisation," *Continuum: Journal of Media and Cultural Studies*, 16: 201–220.

Hollinsworth, D. (1996) "*Narna Tarkendi:* Indigenous performing arts opening cultural doors," *Australian-Canadian Studies*, 14: 55–68.

Howes, C. (2001) "'Emu Tracks' making dreams come true," *Koori Mail*, 17 October: 19.

Howitt, R. (2003) "Indigenous Australian geographies: landscape, property and governance," paper presented at the Geographical Society of New South Wales Conference, "Geography's New Frontiers," March.

Hunter, A. (1996) "Bob Geldof in rock & roll payback!," *On the Street*, 5 August: 12.

Kneafsey, M. (2001) "Rural cultural economy: tourism and social relations," *Annals of Tourism Research*, 28: 762–83.

Leslie, D. and Reimer, S. (1999) "Spatializing commodity chains," *Progress in Human Geography*, 23: 401–20.

MacLeod, G. (2001) "New regionalism reconsidered: globalisation and the remaking of political economic space," *International Journal of Urban and Regional Research*, 25: 804–29.

Manuel P. L. (1993) *Cassette Culture: Popular Music and Technology in North India*, Chicago: Chicago University Press.

Michaels, E. (1986) *The Aboriginal Invention of Television in Central Australia 1982–1986*, Canberra: Australian Institute of Aboriginal Studies.

Mitchell, S. (1996) "Mates, Mabo and Warumpi," *Green Left Weekly*, 24 July: 28.

Myers, F. (2003) *Painting Culture. The Making of an Aboriginal High Art*, Durham, NC: Duke University Press.
Smith, M. (1996) "Tiddas: bridge of voices," *The Drum Media*, 6 August: 20.
Walker, C. (2000) *Buried Country: The Story of Aboriginal Country Music*, Sydney: Pluto.

Index